Green Victorians

Green Victorians

The Simple Life in John Ruskin's Lake District

VICKY ALBRITTON AND
FREDRIK ALBRITTON JONSSON

The University of Chicago Press

CHICAGO AND LONDON

VICKY ALBRITTON has taught at Johns Hopkins University, Colorado State, and the University of Chicago. FREDRIK ALBRITTON JONSSON is associate professor of British history and history of science at the University of Chicago. He is author of *Enlightenment's Frontier: The Scottish Highlands and the Origins of Environmentalism*.

The University of Chicago Press, Chicago 60637
The University of Chicago Press, Ltd., London
© 2016 by The University of Chicago
All rights reserved. Published 2016.
Printed in the United States of America

25 24 23 22 21 20 19 18 17 16   1 2 3 4 5

ISBN-13: 978-0-226-33998-6 (cloth)
ISBN-13: 978-0-226-34004-3 (e-book)
DOI: 10.7208/chicago/9780226340043.001.0001

Library of Congress Cataloging-in-Publication Data

Albritton, Vicky, author.
  Green Victorians : the simple life in John Ruskin's Lake District / Vicky Albritton and Fredrik Albritton Jonsson.
    pages ; cm
  Includes bibliographical references and index.
  ISBN 978-0-226-33998-6 (cloth : alk. paper) — ISBN 978-0-226-34004-3 (e-book) 1. Sustainable living—England—Lake District—History—19th century. 2. Alternative lifestyles—England—Lake District—History—19th century. 3. Cottage industries—England—Lake District—History—19th century. 4. Ruskin, John, 1819-1900—Knowledge—Sustainable living. 5. Ruskin, John, 1819-1900—Homes and haunts—England—Lake District. 6. Lake District (England)—Social life and customs—19th century. 7. Consumption (Economics)—Moral and ethical aspects—England—History—19th century. I. Jonsson, Fredrik Albritton, 1972- author. II. Title.
  GE199.G78A52 2016
  942.7'8081-dc23
                                                            2015029031

♾ This paper meets the requirements of ANSI/NISO Z39.48-1992 (Permanence of Paper).

Contents

Map • vi

INTRODUCTION • 1
Green Victorians

CHAPTER ONE • 21
No Wealth but Life

CHAPTER TWO • 48
Selling Sufficiency

CHAPTER THREE • 70
Queen Susan

CHAPTER FOUR • 96
Taming the Steam Dragon

CHAPTER FIVE • 119
Insatiable Imagination

CHAPTER SIX • 149
Nothing Much

CONCLUSION • 174
Ruskin in the Anthropocene

Acknowledgments • 179
Abbreviations • 183
Notes • 185
Index • 205

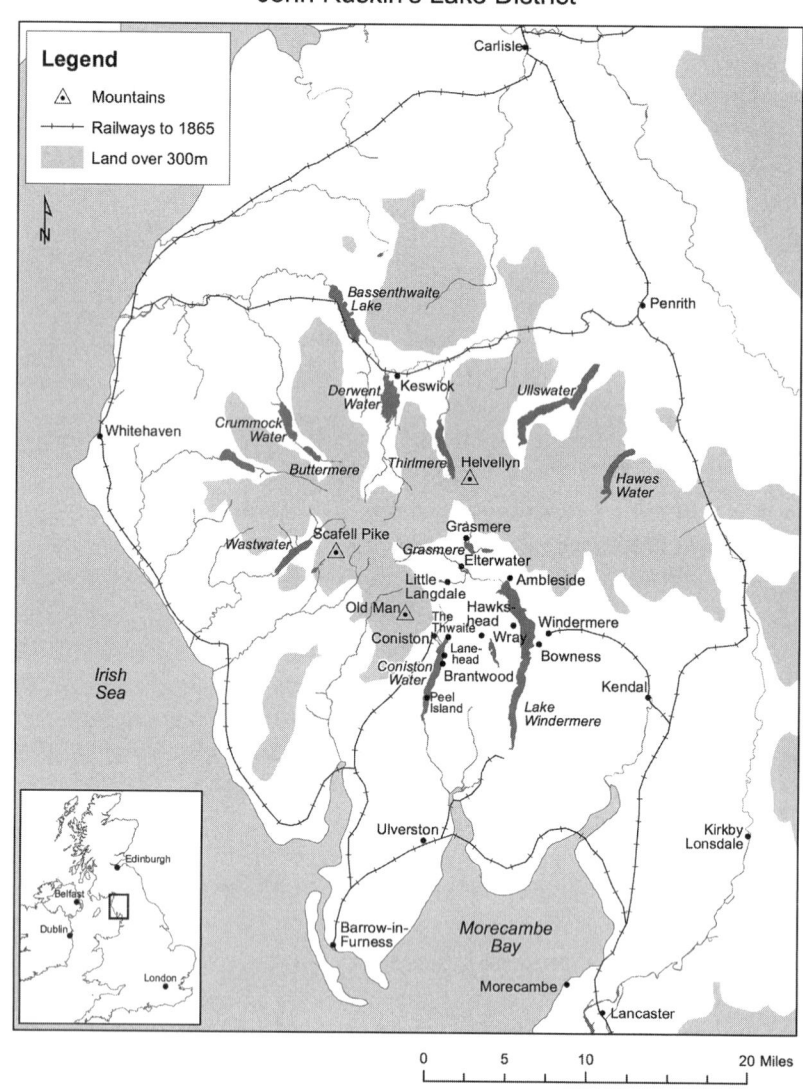

John Ruskin's Lake District

INTRODUCTION

Green Victorians

Sometime in 1883, a London barrister on holiday in England's Lake District entered a small cottage and sat down beside a dilapidated old spinning wheel. Such wheels had fallen out of use with the invention of steam spinning. The middle-aged bachelor was visiting an eighty-six-year-old local woman to learn how to spin. In a charming account of his struggles at the wheel, he complained, "Everything went wrong: the wheel reversed, the thread broke, and the flax twisted itself up into inconceivable bewilderments." Despite these frustrations, he persevered. Soon, with the aid of his housekeeper, he began to teach his hard-won new skills to older local women who could barely remember their grandmothers spinning. In time the barrister, Albert Fleming, became widely known for his rural, community-oriented activities. He reached out to dozens of local women and found one of the few remaining men who knew how to work a hand loom. Together, they began to turn hand-spun yarn and thread into hand-woven linens. Contrary to the cynical predictions of Fleming's London friends, they found they were able to sell their rustic creations for a small, steady profit.[1]

This was deeply gratifying to Fleming. He was appalled that these ancient arts had been eclipsed so rapidly by mechanization. He reflected on the booming industrial world with a bitter cost-benefit analysis: "For 'the virgin labor of her shuttle,'" he wrote, "you shall have cheap Manchester goods; for the sweet singing of poets under blue skies you shall have the roar of ten thousand spindles under black ones." As he saw it, thoughtless consumption had darkened the skies as well as the soul. Spinning had been a stable and satisfying form of work in the Lake District before it was supplanted by steam-powered machines that debased the lives of workers and consumers alike, warping desires and destroying landscapes. No wonder then that Fleming found pleasure in teaching local women skilled handwork. This seemingly innocuous, eccentric community industry was in fact the fruit of a revolutionary perspective. With his band of elderly ladies and his humble cottage factory, Fleming hoped to counter the forward march of mechanized labor. In doing so, he was heeding the writings of the famous art historian and social critic John Ruskin by combining beauty and skilled craftsmanship with joy in work while maintaining an awareness of nature's fragility. In this, Fleming and other followers of Ruskin were attempting to revive and preserve a traditional economy that they associated with a better, simpler way of life.[2]

There could be little doubt about the difficulty of the challenge. Every workday in Victorian Britain, hundreds of thousands of workers were streaming into factories to churn out cheap and often shoddily made furniture, cookware, clothing, and trinkets. Even books were mass produced. In the first half of the nineteenth century, the printing of books had been transformed from a painstaking process of hand printing into a dizzyingly rapid process that made full use of new machinery. Publishing houses rumbled with steam-powered rotary printing presses. The buildings where these presses whirled without cease bore some resemblance to a train station, with wheels spinning, pipes gasping, and pistons chugging. The first steam-powered press had been around for a long time, having been bought by the *Times* in 1814. Most paper in Britain was made by machine by 1830; later innovations included a typecasting machine in 1838, the Wicks rotary caster in 1881, and Linotype machines in 1886. A huge variety of works were constantly coming on the market, including Tennyson's *Tiresias and Other Poems*, 1885 (15,771 copies in one year), Charles Darwin's *The Formation of Vegetable Mould through the Action of Worms*, 1881 (8,500 over three years), Sir John Seeley's *The Expansion of England*, 1883 (80,000 in two years), and Cassell's *Dictionary of Cookery*, 1888 (145,000 in ten years). Such large-scale production offered great hope for the future. New technologies had helped make books and daily newspapers widely available, increasing literacy among the general public. This avalanche of books would seem to have been an unqualified boon for democracy in every sense, helping ease the divide between rich and poor, between the ruling political class and the laborers who could now follow activities in Parliament.[3]

Yet in general, mechanized industries inspired a great deal of ambivalence in Ruskin and his circle of friends and followers. Rapid steam-powered production entailed certain obvious social costs. Skilled handwork was cast aside in favor of dangerous, rattling machines operated by anonymous men and women. Middle-class consumers purchased kettles and footstools with little knowledge of how these items were made. Textile factories in the 1870s roared with deafening machines; spinning belts at times snared women's hair, scalping them. Children were employed to clear textile waste and fiber dust, sometimes falling victim to the machines themselves. Efficiency had become the guiding principle, with great rewards for those who financed and managed the process even as workers suffered. Although the British Parliament sought to curb the worst abuses with factory legislation, new laws were difficult to enforce, leaving many laborers vulnerable to broken machinery, pulmonary disease from dust, cancer from contact with chemicals, deafness from long days in noisy workshops, low wages, and oppressive management. The London Matchgirls Strike of 1888 drew attention to scandalous working conditions in which white phosphorous used for matchstick production caused toothaches, tooth loss, and finally decay of the jawbone. Nor did having work assure anyone of decent living conditions. Housing reformers highlighted the plight of urban slum dwellers, with lurid accounts of gambling, alcoholism, prostitution, and incest in an underworld of filth and despair. In 1885 the journalist W. T. Stead exposed London's child prostitution rings, stunning bourgeois readers who had never set foot in the East End. The darkness of industrial England extended also to the atmosphere. Sooty coal smoke from the factory districts obscured the skies and blighted vegetation.[4]

In the face of so many social ills, critics like Ruskin and Fleming despaired of regulating and reforming the industrial city from the inside. Instead, they withdrew to the countryside in search of a new social order and a more thoughtful approach to the use of nature. A slim book from 1889, *Songs of the Spindle & Legends of the Loom*, opens the door to their alternative economy. A copy of the book survives in the Huntington Library. This modest artifact from Fleming's linen industry presents a concrete example of an experiment in handicraft production that recalled the earliest days of book and cloth making. At over 125 years of age, it is not at first sight an enchanting creation. The dingy beige cloth cover stretches tautly over its sturdy frame and tends toward burlap in texture. The title, the image of a spinning wheel, a partial border, and a flax plant appear in dull cornflower blue paint on the front and back covers. It is a bit worn and the paint is smudged, but clearly the book is intended to call to mind the natural elements of its

FIGURE 0.1 A handmade book: H. H. Warner's *Songs of the Spindle & Legends of the Loom*, 1889.

making; the preface even tells us the linen cover is "unbleached," the "colour of the dried flax," and "the product of hand-work alone." Most pages are well preserved. The crisp, elegant type setting on hand-cut paper suggests thoughtful, high-quality craftsmanship. There are dozens of poems and illustrations, all relating to the theme of hand spinning and weaving, each with the poet's or artist's name beneath it. The first image shows the Langdale Valley in the English Lake District, where the book was made. It is unabashedly picturesque, with undulating fields stretching back toward a modest mountain range. One can imagine in full color the bending gravel roads thick with ferns, the purple heathered hills, and the glassy stream. This image, like everything else about our story, is an invitation to enter the world of its making.[5]

The dream of this handmade book began when Fleming moved to the

FIGURE 0.2 Auto-gravura showing the Langdale Valley, from *Songs of the Spindle & Legends of the Loom*.

Lake District in 1883. He was drawn to the area by a keen desire to be closer to Ruskin, by then a renowned critic of liberal economic thought, well known for his scathing denunciations of mechanized labor and industrial pollution. Ruskin was also the famed author of the five-volume work of art history *Modern Painters* and a key figure for both the Pre-Raphaelite painters and the Arts and Crafts movement. From his first important publication in 1843 until his death at the turn of the century, Ruskin was a major force in Victorian intellectual and cultural life, so highly regarded and influential that he was offered a final resting place in Westminster Abbey (his family declined the privilege since he had already requested a humble burial plot in the Lake District). Fleming's house, known as Neaum Crag, sat in the hills just a few miles from Ruskin's own villa, Brantwood, in Coniston. It thrilled and inspired him to be so close to a much-admired thinker, and soon he was attempting to put Ruskin's theories into practice.[6]

An hour's walk would have taken Fleming from Neaum Crag to the cottage where his weavers, spinners, and paper makers worked on various parts of the handmade book. These country lanes carried him back in time to a society of familiar relations, rural traditions, forgotten skills, and blue skies. This was a world far removed from the squalor and blight of factory towns and urban slums. On a typical day in 1889, he might find Eleanor Heskett and Martha Walker spinning flax for the hand loom weaver John Thursby. The freshly woven cloth was laid out to dry on the hillside, while the paper was folded and sewn by Sarah Coghlan and E. Marshall. Eventually, Fleming assisted the book's editor, H. H. Warner, in compiling a list of all these workers' names and arranging for them to be inserted at the start of the book. We would not know them otherwise. Fleming and Warner intended their book to be a product of local landscape and labor, of skilled hands, mutual respect, and careful husbandry of resources. The linen cover was meant to remind the reader of the hills and grasses; it had after all been bleached "by no deleterious chemicals, but by the pure mountain air and sunshine." It would call to mind with every turn of the page the happy manner of its production. As Warner would write in the preface, "Machine-made goods, with all their superb mechanical finish, are monotonous in their uniformity, and lack . . . human touch, interest, and individuality." But in this case, at least, each of the 250 limited-edition copies would be unique, imperfect, and meaningful. Most books produced by steam power offered only the name of the author, publisher, and printer. Such objects did not convey the story of their creation. The same could be said of any cheaply made bowls or sheets or saucers. By contrast, when a new owner held copy

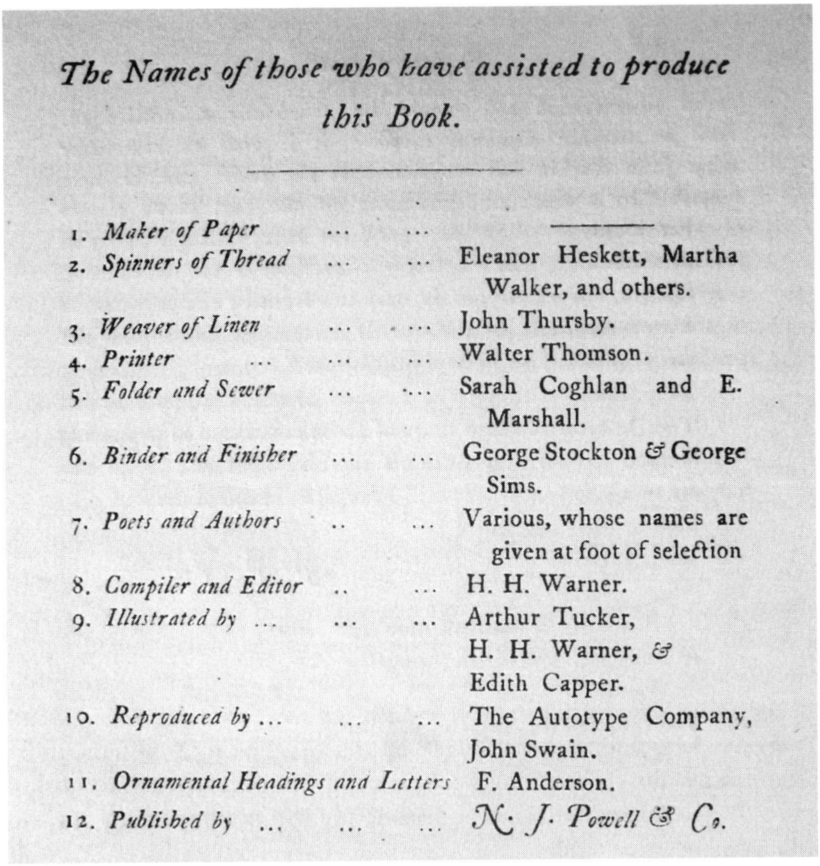

The Names of those who have assisted to produce this Book.

1. Maker of Paper	
2. Spinners of Thread	Eleanor Heskett, Martha Walker, and others.
3. Weaver of Linen	John Thursby.
4. Printer	Walter Thomson.
5. Folder and Sewer	Sarah Coghlan and E. Marshall.
6. Binder and Finisher	George Stockton & George Sims.
7. Poets and Authors	Various, whose names are given at foot of selection
8. Compiler and Editor	H. H. Warner.
9. Illustrated by	Arthur Tucker, H. H. Warner, & Edith Capper.
10. Reproduced by	The Autotype Company, John Swain.
11. Ornamental Headings and Letters	F. Anderson.
12. Published by	N. J. Powell & Co.

FIGURE 0.3 List of all the workers who helped make *Songs of the Spindle & Legends of the Loom.*

#164 in her hands, she could imagine the blue-flowered flax plants gathered from sunny fields, the nimble hands of the woman who spun the flax into fine thread, and the furrowed brow of the weaver leaning over his loom. This is *why* the book was made. The moment of purchase would mean more than just an exchange of money for a commodity; it would illuminate the act of labor itself. In a time when skilled workers found themselves subordinated to the speed and efficiency of mechanical processes, Fleming's little book reminded consumers that there was a different way to think about work and consumption. Tables, kettles, dresses, cups, and books did not have to be produced in circumstances unknown to the consumer.

In recent years, a number of critical voices have called into question the long-term viability of our consumer habits. *Affluenza, The End of the Long Summer, Eaarth, Prosperity without Growth,* and *Plenitude* all argue the necessity of living green in the age of global warming. Forecasts of peak oil, planetary boundaries, ecological overshoot, and anthropogenic climate change challenge the ideology of exponential growth that underpins our politics and everyday life. "We're so used to growth," Bill McKibben observes, "that we can't imagine its alternatives." How can we consume without losing sight of the environment and the welfare of people in distant lands? The path forward, McKibben reckons, starts with a new vocabulary, "words that may help us think usefully about the future. Durable. Sturdy. Stable. Hardy. Robust."[7]

Some readers will wonder how such an ethos could gain any traction with contemporary consumers. McKibben's vision of a simpler life may call to mind an austere and joyless existence, bereft of the myriad pleasures of modern society. All the same, it is something of a truism in the psychology of consumption that an increase in the quantity and variety of goods available to a consumer does little to increase well-being and happiness beyond a certain basic point of comfort. A long line of critics has sought to explore the vexing issue of *how much* might be deemed *sufficient* for any given person. American readers may be familiar with the works of Henry David Thoreau, Helen and Scott Nearing, Wes Jackson, Wendell Berry, and Juliet Schor. In Britain, the issue has been tackled by thinkers from John Ruskin and William Morris to E. F. Schumacher and Tim Jackson. Each of these critics has pursued different philosophical and political paths, but they all share a common commitment to an ethos of artful simplicity, what we will call in this book the *culture of sufficiency*.[8]

Thomas Princen has defined sufficiency as the everyday politics and practice of steering a course between indulgence and abstinence. It is a way of life that makes use of technology and economic growth critically and selectively. Sufficiency is the choice—the "first best" choice—to establish what is enough in everyday . . . use" of technology. This critical view of development is in turn connected to a broader understanding of the relation between human economies and the life-supporting system of the planet. For Juliet Schor, such "an environmentally aware approach to consumption" represents an awakening of the consumer to the "ecological impacts" of production and a desire to "lighten [the] footprint" of consumer spending. Clive Hamilton defines it negatively as the rejection of growth for growth's sake. Consumers in wealthy countries have long suffered from an epidemic of "affluenza"—the psychological and sociological condition "when too much is never enough." Robert and Edward Skidelsky argue that "economic insatiability" is a product of capitalist society. The human propensity "to compare our fortune with that of our fellows and find it wanting" has become a "commonplace of everyday life."[9]

Our book recovers the story of a small circle of men and women in late nineteenth-century England who tried to work out a satisfying practical alternative to mass consumption and industrial society. One hundred fifty years ago, John Ruskin and his followers tackled many of the questions we face today—with illuminating and sometimes disturbing consequences. Their concerns, like ours, arose from a sense that life was becoming less, not more, fulfilling in the age of mass consumption. They were responding to Ruskin's prescient writings on the wealth of life and the Storm Cloud of the Nineteenth Century. Buying less and *buying wisely*, he believed, would pave the way to a healthier society and natural world. This was at once a philosophical and practical experiment, driven by central existential questions about the meaning of the good life. Acts of consumption were inseparable from choices about the public good and the environment. Should one take the train to London to buy the latest, finest linen sheets woven in a factory with power looms, or encourage the revival of hand loom weaving in a nearby village? Should one buy a new dress in the most fashionable color, or stick to clothes with locally sourced natural dyes? This critical approach to consumption brought its own share of headaches and hardships. However, the people in our story showed remarkable persistence and creativity in defending their vision, constantly experimenting with new approaches to a social problem so large and so pervasive. The fruit of their efforts was a culture of sufficiency that covered almost every aspect of life, from food and fuel, art and furniture, to science and the environment.[10]

This quest for the simple life is arguably central to environmentalist ideology, yet few scholars have investigated its historical origins or tried to chart how people in the past attempted to practice this ethos. Most of the research on environmentalism has focused instead on love of the wilderness, strategies of conservation, or anxieties about pollution and overpopulation. The legacy of Enlightenment natural history and Romantic science from Pierre Poivre and Alexander von Humboldt to John Muir has been the subject of numerous studies. Mountain peaks, glaciers, and tropical islands offered these thinkers outdoor laboratories to test new theories of climate, geology, and ecology. Travel journals describing voyages to such spectacular places also helped popularize a quasi-religious love of wilderness. Distant mountains and forests afforded nineteenth-century travelers and readers a space for spirituality. Similar sentiments surfaced in literature and art as well. The Romantic poet William Wordsworth brought the full force and beauty of the natural world to the fore in *The Prelude* (1850). In the famous Pre-Raphaelite painting of drowned *Ophelia* (1851–52), John Everett Millais's reverence for nature appears in the painstaking representation of every single twig, leaf, and petal in the background, the figure of human mortality framed by flourishing vegetation. For these thinkers and artists, nature acquired intrinsic value beyond economic utility or older conceptions of divine design. The campaign to safeguard vast tracts of forested and mountainous regions began with this kind of revolutionary passion and spurred the foundation of the national park system in the United States and other countries. The use of prized areas was soon prohibited; even indigenous peoples were expelled. The ideal of preserving pristine territories inspired readers, artists, and scientists alike.[11]

A second current in the scholarship on environmentalism has focused on national and local practices of conservation, or the sustainable use of land. Historians have explored preindustrial and modern traditions of resource management in forestry, soil husbandry, and fisheries. State building and geopolitical competition compelled rulers to acknowledge and tackle resource problems and ecological bottlenecks from the Venetian Republic to Tokugawa Japan. The problem of governing tropical islands seems in particular to have encouraged this sort of policy. In an island environment, ecological disturbances and resource shortages were both easier to spot and more keenly felt when overexploitation and monocultures triggered crises of production and environmental deterioration. On the island of Mauritius in the eighteenth century, the French naturalist Pierre Poivre pioneered a program of forest preserves in order to thwart a trend of desiccation and climate change, which he attributed to land clearance. Such anxieties were felt elsewhere as well in the late eighteenth century. On the much larger island of Great Britain, population growth and accelerating resource use gave rise to precocious forecasts about the physical limits to economic growth at the end of the Enlightenment. Already in 1789, the mining engineer John Williams warned that the British economy would collapse when the nation ran out of coal. Government officials and naturalists promoted policies of resource husbandry and import substitution. But such initiatives brought authorities into conflict with local communities. Early modern villages and towns struggled to maintain intricate systems of land use set up to protect common resources. In many cases, restricting the exploitation of certain areas intensified competition for the use of the remaining land. The projects of top-down sustainability thus overlapped and conflicted with a widespread system of common-use rights and local resource management.[12]

After World War Two, the debate expanded beyond the question of preservation or conservation. A third current in the scholarship explores the emergence of environmentalism as a potent popular discourse in affluent postwar societies. Land use was no longer just a concern of local communities or national interest; environmental threats began to be conceived of at the global level. In 1948 Fairfield Osborn (*Our Plundered Planet*) and William Vogt (*Road to Survival*) warned that the ills of industrial society threatened the health of the entire biosphere. The onset of the Cold War deepened this fear of planetary apocalypse. Soon, worries about Third World overpopulation gripped American popular consciousness. In the 1968 book *The Population Bomb*, Anne and Paul Erlich predicted mass famine in the near future. E. F. Schumacher questioned the notion that economies could or should grow indefinitely to accommodate burgeoning populations. He argued that a green economy aiming for limited growth offered the only long-term remedy to the problem of strained and declining resources. When the global disaster predicted by the Erlichs failed to materialize, many critics cried foul. Indeed, after 1960, the world population more than doubled and the death rate actually declined. Technological progress, it seemed, had saved the day. Mainstream economists married this faith in techno-fixes to their defense of the power of free markets. On their count, every serious type of scarcity and pollution would trigger a process of innovation and substitution.[13]

Now, in a moment rife with irony, the quarrel between environmentalists and cornucopians (advocates of infinite economic growth) has been rekindled and transformed by the growing evidence of anthropogenic climate change. For the first time, humanity has become a geological agent, capable of reshaping the climate of the planet. Optimists minimize the danger by employing the method of discounting: future generations will be far richer than ours and therefore more capable of dealing with the problem. Another strategy of optimists is to embrace the prospect of geo-engineering. Perhaps we can cool the global climate by adding to the atmosphere new aerosols to counter the greenhouse effect. However, the broader effects of such intervention remain very much in doubt. As Clive Hamilton has shown, scientists are proposing to spray the atmosphere with heat-deflecting sulfur particles, or to stimulate massive algal blooms to suck carbon out of the air, even though no one can say with certainty what the repercussions might be. In general, ecological systems tend to be more complex than we like to imagine, and deleterious unintended consequences all too common. Therefore, the failure to reduce our emissions now could well lead to the crisis spinning out of control. At the same time, the task of transforming the system is nightmarishly complex. If economic growth drives emissions, should we not cut growth to cut emissions? How do we restructure the energy base of our economy without challenging the consumer orientation of our civilization?[14]

This is where Ruskin and his followers claim our attention. Earlier than most critics, they sought a way to exit the fossil fuel economy and consumer society. In a letter from October 1877, Ruskin advised a concerned businessman on how to live a better life. He wrote:

> First. Keep a working man's dress at the office, and always walk home and return in it; so as to be able to put your hand to anything that is useful. Instead of the fashionable vanities of competitive gymnastics, learn common forge work, and to plane and saw well;—then if you find in the city you live in, that everything which human hands and arms are able [to do] . . . is done by machinery,—you will come clearly to understand . . . that . . . to dig coal out of pits to drive dead steam-engines, is an absurdity, waste, and wickedness.[15]

Those who sought a simpler life were to stay active, eschew the encumbrances of fashion, learn physical skills that might be useful, make things by hand, and question the benefits of relying on machine power and the labor of others. Ruskin's message is infused with an ideal of sufficiency quite close in spirit to "maker" fairs and DIY projects. His disgust with coal is not so far removed from our own reaction to news of vanishing glaciers and plastic debris in the ocean. Long before *Silent Spring* and Earth Day, middle-class people worried that consumer society was destroying the natural world.

Though Ruskin's vision has obvious affinities with modern environmentalism, the precise role of his political economy in the making of green ideology has not yet been fully investigated. Some scholars have understood his work to be derivative of William Wordsworth's Romantic vision of beauty and nature. Both men lived in the Lake District, separated by roughly a generation, and both were enamored with the area's mountains and lakes. Yet too much insistence on continuity will obscure the original vision of Ruskin, which linked beauty and work with ethical consumption. Many scholars have commented on Ruskin's disturbing vision of the slum-ridden, money-obsessed Victorian metropolis, but they have usually left

FIGURE 0.4 John Ruskin, ca. 1885. Courtesy of the Brantwood Trust.

aside his understanding of how these markets exploited the environment. Ruskin does receive credit as an early preservationist for his battle against the expansion of the railways in the Lake District, although it is hard for us to escape the verdict that he was fighting the wrong enemy, given that mass transit eventually came to be considered ecologically more sound than automobile travel. Ruskin's place in the history of environmentalism can be seen in other ways too, for instance in his influence on important activists like Canon Hardwicke Rawnsley. In the 1870s, a controversy erupted regarding the construction of a water pipeline and aqueduct between rural Cumbria and the burgeoning city of Manchester. Rawnsley and Ruskin both opposed the damming of Lake Thirlmere. The debate filled the newspapers for months and proved a seminal moment for the environmentalist lobby in Britain, rehearsing a strategy of nature conservancy that would be employed frequently in decades to come. Rawnsley also spearheaded a movement to found the National Trust, an organization dedicated to the preservation of traditional landscapes and architecture in Britain. Rawnsley's National Trust activities and other projects promoted more than just the protection of land; he also wanted to revive and preserve the skills and customs of the common people. Thus Rawnsley and his wife organized local artisan workshops and helped create a market for their products.[16]

What has received a great deal of attention is Ruskin's apocalyptic theology, notably his lament for the spiritual decline of England, its pursuit of mammon, and its brutal alteration of the God-given natural world. These anxieties have been characterized narrowly as the effect of brooding religiosity or declining mental health. It is indeed difficult—as it was for Ruskin's own peers—to disregard his periodic hallucinations, fevers, and depression. Yet too great an emphasis on the apocalypse or madness will distort Ruskin's lasting intellectual contribution, for his most important writings, from *Unto This Last* (1860) to *Fors Clavigera* (1871–84), present a powerful indictment of consumer society's mismanagement of resources. Ruskin became the first great intellectual figure to broach the idea that coal burning gave rise to anthropogenic climate change. This culminated in his 1884 lectures on the Storm Cloud, a bleak vision of planetary degradation. These letters and lectures are discussed at length in chapter 1. Despite the mystical language and obscure symbolism used to convey his sense of moral and physical peril, Ruskin's observations are undeniably astute and unnerving to read. His anxieties about atmospheric pollution eerily foreshadow the dark tenor of environmentalism in the postwar era.[17]

Another way to gauge Ruskin's significance, and account for his undeserved neglect, is to compare him directly to his near contemporary John Muir—the iconic figure of American preservationism. Muir (1838–1914) and Ruskin (1819–1900) shared a Scottish mercantile and Presbyterian background. Both men received an education steeped in Enlightenment natural history and Romantic science, including the writings of Alexander von Humboldt. They traveled widely and loved mountain landscapes, in Muir's case Alaska and the Sierra Nevada range of California, and for Ruskin the Lake District and Switzerland. Both developed a strong fascination with glaciers, to the point of engaging actively in contemporary controversies over glacial science. In middle age, each man settled on a large piece of property, Muir on the Strentzel ranch outside San Francisco, Ruskin at Brantwood near Coniston. They were both keen to shape public opinion on social and environmental issues, though Ruskin shied away from direct engagement with politicians. What separated them was a philosophical difference on the meaning of natural beauty.[18]

Ruskin was by training and inclination steeped in the fine arts, classical learning, and the history of landscape painting. The artistic representations of nature with which he was most familiar included pastoral scenes with shepherds or mythological beings, gentle lakes and hills. At the same time, he championed the work of J. M. W. Turner, whose innovative landscapes often depicted the fiercer, more spectacular elements of nature. But Ruskin tended to appreciate Turner's understanding of the human element in the midst of nature—the smoke-filled urban scenes, the train roaring through its own steam, a barely visible mast engulfed by a roiling ocean. Ruskin memorably praised Turner's painting *The Slave Ship* (1840), in which a frightening storm at sea takes up a majority of the canvas. Yet the sea is not the only subject; instead, the tiny ship with its ill-fated passengers suggests an anthropocentric slant. Ruskin wrote that the ship was "guilty" and "girded with condemnation in that fearful hue which signs the sky with horror." The storm's ferocity was filtered through the human gaze.[19]

John Muir's experience of Half Dome in Yosemite National Park in 1875 could not have been more different. To his mind humans added nothing of worth to the mountain. In fact, human influence would taint its majesty—if only temporarily. He lamented the failure to keep Half Dome cordoned off from hikers: "Now the pines will be carved with the initials of Smith and Jones, and the gardens strewn with tin cans and bottles." But he took solace in the mountain's power to slough off any human debris: "the winter gales will blow most of this rubbish away, and avalanches may strip off the ladders . . . When a mountain is climbed it is said to be conquered—as well say a man is conquered when a fly lights on his head." He continued, "Tissiack will hardly be more conquered, now that man is added to her list of visitors. His louder scream and heavier scrambling will not stir a line of her countenance." In Muir's world, nature was the most wondrous when human influence was removed, whereas Ruskin was never quite at ease with the idea of pure, uninhabited wilderness.[20]

A passage from *The Seven Lamps of Architecture* (1849) illuminates Ruskin's aversion to an unpeopled natural world still more, while clarifying what he felt humans brought to nature. He recounted a walk in the forest among the subalpine mountains of Jura: "[C]lear green streams" crossed the valley in a bed of oxalis and wood anemone, with a hawk sailing in the sky high above the long-inhabited foothills. It was a scene of "secluded and serious beauty." Then a strange thought suddenly cast a "blankness and chill" over the landscape. He imagined that it was a "scene in some aboriginal forest of the New Continent." In an instant, the lovely forest transformed itself from a blissfully secluded idyll to a *never*-peopled world, a natural order devoid of human influence. Such a place could hold for Ruskin none of the beauty

FIGURE 0.5 The Old Man of Coniston seen from the Brantwood lawn, 2011.

of tamed woodland "dyed by the deep colours of human endurance, valour, and virtue." He pulled himself back from the unnerving reverie, reminded that on *this* horizon, thankfully, were familiar marks of civilization and architecture—"the iron walls of Joux, and the four-square keep of Granson." For Ruskin, only through human memory and art did the wilderness shed its alien countenance and gain meaning and value. It was the "union of story and scenery," the ability of humans to weave their own narrative through the forests and mountains and plains, which created beauty, not nature in its own right. To Muir, this was a mistaken view. All of nature was sacred and marvelous, whether touched by humans or not. Such "holiness" reached its highest pitch among the lofty peaks rather than the lowlands, along a vast and barren cliff face rather than a well-worn foot path. Muir wrote to Ralph Waldo Emerson: "I do not beckon you because mountains are more glorious than plains, but because they are less glorious, because they are simple & absorbable. Here we may most easily see God." At the heart of Muir's preservationism, then, lay the alien presence that sent chills down Ruskin's spine.[21]

While there is much to admire in Muir's conception of nature, it glossed over the mundane realities of human existence: agriculture, shelter, work, and population. This contrast, more than any other reason, may explain why Ruskin has been neglected by historians and scholars of environmentalism. For those who place the cult of the wilderness at the center of environmentalist thought, Ruskin is, no doubt, like Emerson, too much a denizen of the lowlands, too much concerned with human welfare and common activities. But we might just as well turn the criticism on its head. In the age of anthropogenic climate change, environmentalist thought has to deal first and foremost with the problem of consumer demand and carbon emissions. And here, Ruskin may be of greater relevance than Muir. Ruskin believed the way to a more fulfilling life required a critique of classical political economy; the laws of supply and demand had to be seen within a far more expansive framework that included foresight regarding the use of natural resources. At the heart of this critique was an ethics of consumption—summed up by the famous dictum "There is no wealth but life." In Ruskin's view, the landscapes of the Jura and the Lake District were shrouded in beauty and poetry. But the fate of the natural world depended on the place of ethics in the economy. For every purchase might support another factory, another lost forest, another smokestack.[22]

Ruskin's practical efforts to live a simpler life in England's Lake District inspired his followers as much as his theory of labor and virtue. The story of Ruskin's move north is the subject of our first chapter. At his home, Brantwood, overlooking Coniston Water, he set in motion a series of projects, exploring the possibilities of sufficiency. He laid out a garden to determine what a working man might be able to manage if he had his own allotment.

He studied native plants and shunned greenhouse exotics. He collected geological specimens to display in local schools so that children might develop a deeper connection to the landscape. Physical labor and exercise were moral aims in themselves for Ruskin. Often he simply donned a pair of rugged trousers, went outside, and worked with his own hands. At a time when wealthy men employed housefuls of servants (as he did himself), it was something of a marvel to see an Oxford professor chopping wood. Another area of practical activity for Ruskin was philanthropy. He donated a great deal of his own money to establish the Guild of St. George, a working-man's organization that aimed to enrich the lives of laborers, shopkeepers, and any other workers without much formal education through artistic workshops and community gardening. In all of these projects, Ruskin sought to teach people about the natural world, lure them away from conventional consumer habits, and broaden the idea of the good life.

The story of Fleming's Langdale Linen Industry, mentioned earlier, is told in chapter 2. Through his and his coworkers' determination and perseverance, the small business began in 1883 and ultimately achieved modest financial success, proving profitable in different guises until the 1920s. Fleming was far from alone in following Ruskin. *Green Victorians* examines the lives of a whole cast of unlikely characters. The charismatic Coniston gardener Susanna Beever—the "Owl of the Thwaite," as she was known by her admirers—represented for Ruskin a profound spiritual connection to both plants and animals. In chapter 3 of the book, we explore her gentle type of idealism, informed by an arcadian fascination with local flora, customs, and landscapes. Visitors flocked to see her gardens and enjoy tea with the lady who, for Ruskin, embodied the virtues of sufficient living. But it was a lonely business to practice Ruskin's ethics in a provincial backwater. Beever found herself a helpless eyewitness to Ruskin's descent into madness. During the most tumultuous of Ruskin's manias, Albert Fleming's friendship and his revival of the spinning industry provided a measure of comfort. Fleming edited *Hortus Inclusus*, a selection made from Ruskin's 900 letters to Beever. Beever's correspondence reveals an unusual ability to turn everyday life into a world of wonder and delight.

Chapter 4 covers Canon Hardwicke Rawnsley's work, which included much more than founding the National Trust and protesting railways. Charismatic, ambitious, and more than a little arrogant, Rawnsley was very capable of winning support for his causes among locals and visitors alike. In parallel with Fleming's Langdale linen project, Rawnsley developed a local Arts and Crafts workshop, known as the Keswick School of the Industrial

Arts. This institution gave laborers new skills and nourished the creativity of men and women who might otherwise not have known the pleasure of artistic imagination. The KSIA achieved solid financial success as well. Rawnsley even found a student and ally in none other than Beatrix Potter, the children's author, one of the most famous contributors to the National Trust and a hands-on Lake District farmer for the second half of her life. Despite such successes, Rawnsley was troubled by Ruskin's bleak forecast about pollution—likely taking his cue from the letters in *Fors Clavigera*. He wrestled with the foul effects of coal smoke on humans and nature alike. But he also listened, perhaps too respectfully, to the factory lobby and its proposals for new, sometimes dubious smoke-abatement technologies.[23]

Chapter 5 introduces William Gershom Collingwood, who had been Ruskin's student at Oxford and had a bright academic career ahead of him. Yet he abandoned a life of relative security and steady work to become Ruskin's private secretary in the Lake District. In addition to editing Ruskin's writings, he became a painter, championed amateur archaeology, and wrote historical romances about the virtues of the simple life. He later published a tourist guide to the Lakes that envisioned the landscape as a thriving post-industrial society. Life for Collingwood, his wife, and his four children was sometimes financially precarious, though they still managed to fill their house on Coniston Water with a wealth of art, music, and custom-made furniture. Out of all the figures studied here, no one else more clearly sacrificed an otherwise lucrative career in the city in favor of an alternative way of life. In this way, Collingwood's story offers a blueprint of sorts for future experiments in simple living.

Not all of those who followed Ruskin consciously chose to do so. In chapter 6, we turn to the lives of Collingwood's four children to catch a glimpse of what it is like to experience sufficiency daily within the home—from the moment of birth. Even when very young, these children were surrounded by music, art, and literature. They roamed through the landscape of Coniston, learning natural history in the field. The children were strange and aloof, but restlessly creative, making their own amusements. They produced a family magazine called *Nothing Much*, which was circulated privately to a small list of subscribers. This showcased their precocious talents as artists and writers. There were stories of rambles and sailing escapades, along with natural history, archaeology, fiction, and whimsy. The children even made mock advertisements for their magazine, developing a keen appreciation of how consumer desires were shaped by the forces of the marketplace.

Collingwood's cheerful and artistic family life made a deep impression on his "adopted" son, Arthur Ransome. We argue that Ransome's much-loved children's book *Swallows and Amazons* (1930), a story about the elaborate games played by a group of children on the water and islands of the Lake District, was in fact a gently disguised account of his life with the Collingwood family. In Ransome's book, a great deal of attention is given to practical skills, the power of accurate observation, and imaginative play. The children in the book are modeled more or less on Collingwood's grandchildren, who learned these skills and habits from his daughter Dora. Through Ransome's books and their multiple television and film adaptations, several generations of British children have imbibed a popular version of Arts and Crafts culture without even realizing it.

By experimenting with the fabric of their own lives, Ruskin and his followers sought to redefine the meaning of wealth. In both word and deed, they tried to show that the good life depended less on material abundance and social prestige than on artful simplicity and the bonds of family and community. For this reason, our story is as much about things and places as it is about people. The social and aesthetic impulse of Ruskin's ideal was richly manifested in the quality of the objects it produced, from linen and wood carvings to tourist guides and children's magazines. The peculiar ecology and land use of the Lake District also played a major role in shaping the values of the movement. Ruskin and his followers articulated their own kind of preservationism dedicated at the same time to protecting the landscape and the traditions of its people.[24]

This double mission to preserve the land and its customs was perhaps the Achilles heel of the whole enterprise. Although Ruskin and his allies succeeded in establishing a thriving handicraft industry and protecting the Lake District from overdevelopment, they failed in the larger task of creating a long-lasting alternative to industrial society. Sufficiency proved insufficient. As we shall see, there were many reasons their experiment did not leave a lasting mark on British society. The aim of this story is not to smooth over the contradictions and failures of the movement, but to give them a central place in our account. This is the best way to take seriously the ideals at the heart of Ruskin's Lake District community.

CHAPTER ONE

No Wealth but Life

In 1887 a well-to-do sixty-eight-year-old gentleman suffered a fit of insanity and went on a shopping spree. He traveled with his valet to the southern coast of England in Kent and took up residence at a hotel in a fashionable district overlooking the ocean. The protective and wary valet called in a hairdresser so that his master would strike a suitably modern figure at the Pavillion Coffee Room, where he chose to dine. The cut did not please the gentleman in the least, and he angrily sent the hairdresser away. He next had "two new Top Coats made," one flamboyantly long, the other short, and several "fine waistcoats." Thereafter, he bought a toy boat and some "guards for his windows." The valet uneasily deemed these purchases "little extravagances but of no account" in his daily letters to his master's cousin. Yet there was more. The gentleman was intrigued by the hotel's assortment of champagne, and ordered his valet to "bring up three [bottles] so that he might study the labels." In a cavalier gesture, he gave away two to the maids and half a sovereign with some sherry to the cook and another servant. The valet complained in another letter that his master had already given three pounds to one of the housemaids. Indeed, he had taken a fancy to her, bought her an outfit, and requisitioned her as a personal servant. The valet was mortified. It seemed the whole hotel was now whispering that his master was "a good bit queer."[1]

For other men of the same class, this kind of spending would have been excessive. But for the man in the Pavillion Coffee Room, the shopping spree was a profound personal calamity. The nature of the failure was, strictly speaking, not material. Though his liquid assets had been dwindling for years, he still owned two large houses and enough fine art to fill a museum. There was no chance that he would end his life in a poorhouse. No, the disaster here was moral rather than financial. In the increasingly agitated letters home, the valet chronicled the breakdown of an extraordinarily sensitive and selfless character. The gentleman, John Ruskin, renowned author and the former Oxford Slade Professor of Art, had succumbed to a chronic illness that involved monthlong attacks of mental derangement, hallucinations, delusions, and paranoia. Joan Severn, his caretaker and cousin, was the long-suffering eyewitness to such altered states. She had also watched Ruskin's estate shrink yearly, thanks to his exceeding generosity among friends and fellow travelers. Just a few months earlier, he had given a former housemaid a check for £325. Normally, lifelong principles informed Ruskin's expenditures and charity; even if Joan wished he would tighten the purse strings, she could see that he was acting in a sound state of mind. Now, in Folkestone, he appeared to have fallen into the habit of making purchases without thought or care.[2]

The situation was one of peculiar poignancy. Ruskin had devoted the past few decades to articulating the need for consumers to spend their money more wisely, blasting the excesses of nineteenth century affluence, and warning of environmental degradation caused by coal-powered factories. What may have seemed like eccentric behavior for a wealthy older bachelor struck the loyal valet, Peter Baxter, as part and parcel of a profound mental crisis. Joan Severn, the valet, and the former housemaid all knew Ruskin's behavior was aberrant (the housemaid had shown the £325 check to Joan, and was told Ruskin's personal account could not have covered it in any case). Ruskin had never hesitated to buy fine works of art, rare books, good works, necessary items, or anything else his cultured scruples deemed worthy. As for clothing, when residing at Brantwood, he had taken to "the more comfortable dress of the homespun country squire" as a sign of his support for honest labor and fair wages. As far as unnecessary luxuries went, he urged restraint, telling his readers that

> The most helpful and sacred work which can at present be done for humanity, is to teach people . . . not how "to better themselves," but how to "satisfy themselves." It is the curse of every evil nature and evil creature to eat and *not* be satisfied.

Ruskin pondered this idea of sufficiency, or as he would have termed it, satisfaction, amid the astonishing excesses of the midcentury "Retail Revolution." Middle-class people were able to buy ever-increasing numbers of gadgets, garments, delicacies, and entertainments. This was the age of the first department stores. The five floors of Maple's on Tottenham Court Road in London contained a storefront that "stretched the length of twenty-five houses" and "employed about 2,000 people." Consumers strolled down its vast aisles to peer into "recessed rooms, complete with windows draped in rose-tinted fabric, and furniture of all sorts arranged in room settings." The warehouse alone "was the size of a modern townlet, encompassing the space of fifty-two ordinary dwellings." It was this new world of delight and pleasure that Ruskin had come to despise. Men and women had grown accustomed to purchasing newly affordable furnishings, textiles, out-of-season produce, and rich meats. These modern-day "Caligulas" had glutted themselves on luxuries, and now they could enjoy only the sort of pleasure that condensed "the labour of a million of lives into the sensation of an hour."[3]

Ruskin won a small but devoted following with his criticism of industrial society, which involved one problem above all: how much should a person consume? This simple question took hold of Ruskin's imagination for the latter half of his life. It shaped his critique of industrial working conditions and inspired a conception of the natural world that may sound familiar today. Current catchwords of the modern environmental movement, like *sustainability*, *sufficiency*, and *plenitude*, all turn on this problem. Ruskin articulated the promise and difficulty of cultivating sustainable consumption. He wrote in *Unto This Last* that the value of a thing depended on the ability to know and use it well: "A truly valuable or availing thing is that which leads to life with its whole strength." For Ruskin, wise consumption combined ideals of social justice, humanist technology, and environmental stewardship. To consume well was to promote the dignity of skilled labor, to further the beauty of well-made designs, and to preserve the natural world from waste and degradation. Crucially, Ruskin recognized that such limits to consumption were not only morally necessary but also a means of enriching life and making it more meaningful.[4]

He reached out to his readers far and wide in a newsletter, *Fors Clavigera* (1871–84), written for the benefit of the "workmen and laborers of Great Britain." But it was his local followers near Brantwood—a diverse group with interests in preservationism, landscape painting, natural history, archaeology, and artisanal work—that came under his influence most directly and intimately. Between 1880 and 1920 they fostered a thriving cultural scene and handicraft industry known as the Lakeland arts revival. For Ruskin and his friends, this social experiment offered a taste of the postindustrial future. Ruskin urged his followers to look forward to a "sweet spring-time"

FIGURE 1.1 The Ruskin household out on the ice near Brantwood, during the Great Freeze of 1895. From left to right: George Usher (Coniston builder who worked on Brantwood), Dawson Herdson (head gardener since 1871), John Ruskin, Joe Wilkinson (under-gardener, later head gardener), Joan Severn, Samuel Clarke (neighbor), Lily Severn (daughter of Joan and Arthur Severn), Charlie Baxter (age 14), Jimmie Baxter (age 9), Peter Baxter (on the tricycle, age 50, Ruskin's valet since 1876, father of Charlie and Jimmie next to him), John Wilson (age 21, stable boy, then coachman). Photograph by John McClelland of Liskard. By kind permission of the Ruskin Museum, Yewdale Road, Coniston, Cumbria, LA21 8DU, UK.

for "our children's children . . . when their coals are burnt out, and they begin to understand that coals are not the source of all power Divine and human." Ruskin's prediction echoed the forecast made by William Stanley Jevons in *The Coal Question* of 1865. The political economist calculated that British coal production would soon reach its highest level of extraction. In fact, output peaked right before the Great War at 287 billion tons. Ruskin's arts and crafts community in the Lake District thus sought to anticipate the social condition that he believed would follow the exhaustion of the British coal mines. Economic contraction and resource scarcity—the end of fossil fuel growth—would offer an extraordinary opportunity to reject the structures of industrial capitalism and create a better society founded on skilled labor and the pleasures of the simple life rather than profit and competition. The ethics of consumption, then, by its very nature, bridged the gap between theory and practice. By delving into Ruskin's critical writings, lectures, and even his diaries, we will see how his sense of environmental fragility and his concern over consumer excess were woven into the fabric of his biography and community.[5]

How exactly did Ruskin come to argue so forcefully for these ideas? Who were his interlocutors, and how were his arguments received? To what extent did he abide by his own ideals—and how did he wind up at Folkestone in 1887, spending so heedlessly that he appalled his friends and family? To answer these questions, we must attend not only to personal biography but also to questions of the household, which meant for Ruskin a combination of family, labor, and natural resources. His move in the 1870s to the northern countryside is especially important, since it promised a sanctuary from the material desires and industrial blight of the Victorian city while offering a space for experimentation. At his house, Brantwood, overlooking Coniston Water and the Old Man of Coniston—the great hill to the west of the village—Ruskin launched a moral crusade against mass consumption and urban life.

THE BIRTH OF A CRITIC

Ruskin's father, John James Ruskin, was of Scottish heritage, while his mother, Margaret, was English. She was also a cousin, and the companion of John James's mother. The couple endured a lengthy engagement as John James sorted out his father's debts. Consequently they were married late and had only one child. John Ruskin was born to them on February 8, 1819, in Brunswick Square, London. John James eventually accumulated a sizable fortune importing sherry and wine. Their son grew up much loved, not to say coddled and encouraged, by his ambitious parents, who catered to the precocious child's every need. Young Ruskin took to this hothouse atmosphere rather well, learning to read by the age of four and showing great talent at drawing and painting. Already as a boy, he sketched the landscapes around him with startling accuracy and a mature touch. He also wrote poems from a very early age, including an encomium to the Lake District at the age of nine in "On Skiddaw and Derwent Water"; he published it at age eleven. The family soon moved to Herne Hill in South London, where the young Ruskin enjoyed a lush, green, wide-open landscape with views of the Thames Valley and the forested Norwood hills. This childhood Eden would play a significant role in his evolving conception of true wealth, especially as the land fell prey to suburban expansion and was subdivided over the next few decades.[6]

Ruskin's evangelical mother had been raised with a stern Presbyterian taste for Scripture, and as such she took it upon herself to read the entire King James Bible with her son. When they were finished, she repeated the exercise. The Bible thus had a lasting impact on Ruskin's education, though it extended as well to Shakespeare, the *Iliad*, Sir Walter Scott, and (at the age of ten) Adam Smith's *Theory of Moral Sentiments*. His parents hired private tutors and traveled extensively, all the while grooming him for life as a clergyman or poet. His early passions for art, literature, poetry, and religion culminated in the sort of erudition and refinement to which his parents aspired. He went up to Oxford in 1836. There he worked hard and won praise, though his health suffered under the strain. Nevertheless, in 1843 at the age of twenty-four, Ruskin published—anonymously—the first volume of *Modern Painters*. Few books were sold initially, but it quickly won the admiration of William Wordsworth. By the autumn of 1846, although his name had yet to be printed in his books, people in the art world were beginning to take notice. His reputation as an art critic was established. *Modern Painters* was a spirited defense of the work of J. M. W. Turner, but also a veritable font of aesthetic, architectural, social, and moral theorizing that would continue to widen in scope and evolve in its aims throughout the subsequent volumes. He might have continued pursuing art criticism, teaching at Oxford, mingling with the establishment—this would have pleased his parents. Instead, he slowly but surely took up the mantle of social critic.[7]

The reasons for his change of heart are perhaps not difficult to understand. Throughout Ruskin's childhood, British industrial centers were growing prodigiously and employing ever more workers in large and dangerous factories. There were few regulations in place to safeguard adult workers, and none at all for children, who were thought to be cheaper to employ and able to do small-scale, nimble work. In 1833 the first Factory Act was passed, prohibiting children under the age of nine from working in textile factories—the same year the British Empire abolished slavery. When Ruskin went up to Oxford, then, such social questions were entering public opinion and lawmakers had begun to take action. The Factories Act of 1847 further restricted working hours to ten per day for women and children. The potato blight and ensuing famine in Ireland spurred more discussions of what was to be done for those who could not help themselves. Such a social awakening among the British public affected Ruskin too, as he began to question the basic understanding of consumer demand and the underlying structures of industrial society.[8]

VICTORIAN HOME ECONOMICS

While concerns about factory work and its wasteful products formed a central current in Ruskin's thought, so too did its increasingly devastating effect on the natural world. In middle age Ruskin would often recall the countryside around his childhood home near Addington. This had once been "a quite secluded district of field and wood, traversed here and there by winding lanes." Interspersed through the landscape were the houses of the gentry together with the "cottages or small farmsteads" of the rural laboring classes. Ruskin dwelled lovingly on the look of the cottages: "wood-built usually, and thatched, their porches embroidered with honeysuckle, and their gardens with daisies." They were "all neatly kept, and vivid with a sense of the quiet energies of their contented tenants." But in the present day, "that same district" was "covered by . . . many thousands of houses built within the last ten years, of rotten brick, with various iron devices to hold it together." There seemed to be a direct correspondence between a loss of aesthetic taste and a dismissive attitude toward the environment.[9]

Ruskin took special notice of the identical "parallelograms of garden" attached to every pair of houses, "laid out in new gravel and scanty turf." This unlovely monotony was entirely representative of the "'rising' middle classes about London." The overwhelming majority of men and women living there had no "common" "skill, knowledge, or means of happiness": "Not a member of the whole family can, in general, cook, sweep, knock in a nail, drive a stake, or spin a thread." They were even less capable of "finer work" like "painting, sculpture, or architecture." They read nothing of value, never thought "of taking a walk," and did not "enjoy their gardens, for they have neither sense nor strength enough to work in them." Instead, their lives were filled with cheap thrills: "The women and girls have no pleasures but in calling on each other in false hair, cheap dresses of gaudy stuff, machine made, and high-heeled boots, of which the pattern was set to them by Parisian prostitutes of the lowest order." The men were no better, with "no faculty beyond that of cheating in business; no pleasures but in smoking or eating; and no ideas . . . of anything that has yet been done of great, or seen of good, in this world." To Ruskin, this new suburban civilization was rootless and transient. The dwellings were so poorly made and little loved that he hesitated to call them inhabited houses. At best, these people were mere "lodgers in . . . damp shells of brick." Such houses were "packing-cases" for temporary storage. The lack of concern for natural beauty led to a vicious downward spiral. The people grew more alienated from the spiritual meaning of the natural world the more it came to be defaced by the man-made structures of suburbia.[10]

Having witnessed such a jarring transformation in the childhood landscape he loved, Ruskin began to formulate his theory as to the causes of this great change—but it was no easy task. He did not mean to appeal merely to lovers of nature or the pious who believed in protecting God's creation. He was determined to contend directly with the most influential liberal economic theorists. Much of Ruskin's social vision was in fact articulated in direct opposition to John Stuart Mill, Adam Smith, and David Ricardo. Again and again in Ruskin's writings, he attacked the assumptions underpinning classical political economy. Human beings did not live for profits alone, according to Ruskin. Free trade and industrial production did not present the road to universal prosperity but instead led to the cruel exploitation of the poor and empty riches for the elite. There was nothing natural or rational about such an economy. Ruskin spoke bitterly of the suburban middle class as a new grotesque species, "monkeys that have lost the use of their legs" and lived only to imitate each other in a caricature of human potential. The advanced stage of specialization, praised by the liberal economists as a crucial key to modern productivity, was for Ruskin an unnatural division between the hand and the mind, which degraded "the operative into a machine." This worry had been voiced before, notably by Adam Smith himself. In Smith's famous parable of the pin manufacture, workers became more effective by endlessly repeating the same simple tasks. He recognized that manufacturing of this sort could stunt the mind and body of the operatives, and recommended public education as a counterweight to the deadening effect of mechanical labor. Workers who received training in arithmetic and reading, Smith thought, would be less susceptible to mental and physical torpor.[11]

Ruskin's remedy was far more radical. Against the reality of factory specialization, he held up the work of the medieval artisan as the model of authentic labor and the practical solution to a multitude of contemporary problems. Ruskin's great study of architecture, *The Stones of Venice* (1853–55), had moved his thought in this direction. The purpose of labor was the skillful making of useful and beautiful things. In *Unto This Last*, Ruskin defined the proper task of political economy precisely in these terms. It was not an abstract science about the interplay of self-interest and markets but a practical guide to a life lived well:

> Political economy (the economy of a State, or of citizens) consists simply in the production, preservation, and distribution, at the fittest time and place, of useful and pleasurable things. The farmer who cuts his hay at the right time; the shipwright who drives his bolts well home in sound wood; the builder who lays good bricks in well-tempered mortar; the housewife who takes care of her furniture in the parlour and guards against all waste in her kitchen; and the singer who rightly disciplines, and never over strains her voice, are all political economists in the true and final sense: adding continually to the riches and well-being of the nation to which they belong.[12]

For Ruskin, political economy was an inclusive art, incorporating the mind and the body, wants and needs, agriculture and arts, women and men. The housewife's prowess with provisioning was as important as male work outside the home. Food, shelter, and music were given equal weight. Good work avoided waste and promoted economy. All labor required knowledge of nature and the body: how to drive a nail into wood, when to sow grain, how to care for one's voice. This striving for individual excellence was also a social enterprise: every laboring person had a specific role to play in making the common good.

This practical orientation of political economy was never far from Ruskin's mind. Soon after completing his critique of liberal economics in the 1860s, Ruskin began to contemplate how the industrial economy might be transformed in practice. In 1871 he bought a large hillside property in the Lake District. Brantwood, as it was called, was situated about a quarter of a mile up from Coniston Water and possessed striking panoramic views with sheep wandering the fields below. Behind the house was a rough terrain with small streams and thick coppice woods. By this time, he had suffered the humiliating annulment of his marriage to Effie Gray (on the grounds of non-consummation) and the loss of both his parents. The move to Brantwood, however, was not simply a form of escapism nor, as one biographer has suggested, an effort to reconnect with his lost childhood. Brantwood was an experiment as much as a retreat. Ruskin went north in search of the good life. There he threw himself into his rural pursuits, alternately engaging in physical and theoretical work—the two inextricably linked, by design.[13]

Ruskin's guide in much of this was the ancient philosopher Xenophon. In the summer of 1875, he commissioned two of his students from Oxford— W. G. Collingwood and Alexander Wedderburn—to come to Brantwood and translate Xenophon's *Economist* into plain English for "British peasants." The text was to make up part of a range of volumes called the Biblioteca Pastorum. The two young men had already demonstrated their devotion to Ruskin at Oxford by taking part in the famous "Hinksey Dig"—one of

FIGURE 1.2 Brantwood, showing Ruskin's lookout tower and Coniston Water, before 1905. Courtesy of the Brantwood Trust.

Ruskin's social experiments—in which wealthy young Oxford scholars undertook to build a road for a poor community with their own hands. The point was not merely to give the young men a lesson in social work, but to stress the benefits of physical labor. Ruskin feared that modern men were losing touch with themselves and the land. The new translation of Xenophon brought home the mutual relationship between these intellectual and physical labors. Fittingly, Collingwood and Wedderburn spent mornings at Brantwood rendering Greek into English, while in the afternoons they helped Ruskin build a dock on the beach below Brantwood. *The Economist* was in part a practical manual for household management. Good wives must be taught to economize, not to buy frivolous things, not to wear makeup, not to eat to excess. They must make sure that domestic servants maintained a humble and restrained lifestyle as well: "you will have to take charge of everything that is brought into the house, distributing it when wanted, and providently taking care of the stores." Prudent management of needs and wants eliminated waste and excess, "so that we may not consume in a month what was meant to last a year." It is a good guess that Ruskin wanted Xenophon's book to guide the household at Brantwood, though in practice he had to cede power to Joan Severn. She was his caretaker and the manager of the house, but she had a family of her own and often imposed her own notions of propriety and economy.[14]

Ruskin used Xenophon's ancient writings as a counterpoint to fashionable political economic treatises of the day. He was also indebted to certain liberal thinkers, perhaps most importantly J. S. Mill, who asked in *Principles of Political Economy* how society might come to terms with permanent physical limits to economic growth. What would happen if the population of Britain grew too large for the land to support it? Unlike earlier economic thinkers such as Malthus, Mill's answer was surprisingly optimistic. Humanity should embrace the virtues of this stationary state (the stage of history when economic development reached its physical limits) long before the physical limits to growth had become pressing and severe. This voluntarily stationary society was free to redirect its fundamental creative urges in new directions: "There would be as much scope as ever for all kinds of mental

FIGURE 1.3 Brantwood harbor ca. 1876, with the dock, which W. G. Collingwood and Alexander Wedderburn built while translating Xenophon's *Economist*. The women's identities are unknown. From W. G. Collingwood, *Ruskin Relics*.

culture, and moral and social progress; as much room for improving the Art of Living, and much more likelihood of its being improved, when minds ceased to be engrossed by the art of getting on." While there is little indication that Mill sought to build a new social order on the basis of his idea, it seems that Ruskin had precisely such a goal in mind.[15]

He took issue with the abstract understanding of work and desire put forth by the liberal economists. Political economy, according to Ruskin, was closely linked to questions of taste and the fine arts. A wise consumer should recognize the skill and labor invested in every object. Consumption required knowledge not only of the commodity but also of the social relations involved in production: "In all buying, consider, first, what condition of existence you cause in the producers of what you buy; secondly, whether the sum you have paid is just to the producer, and in due portion, lodged in his hands."[16] The producer in turn should limit his production to those goods that were genuinely useful and wanted by the consumer. In this way, Ruskin rejected the conventional economic model of Smith and Mill, in which mass production of cheap goods fueled endless demand. Ruskin made it clear that true value was measured by durable, handmade goods instead of mass-produced and disposable commodities: "The value of a thing, therefore is independent of opinion, and of quantity." Ruskin believed that "intrinsic value is the absolute power of anything to support life." This was for him a combination of biological necessity and aesthetic value: "A sheaf of wheat of given quality and weight has in it a measurable power of sustaining the substance of the body; a cubic foot of pure air, a fixed power of sustaining its warmth; and a cluster of flowers of given beauty, a fixed power of enlivening or animating the senses and heart." Ruskin recognized that most kinds of consumption did not fulfill such basic needs. As he put it: "Three fourths of the demands existing in the world are romantic; founded on visions, idealisms, hopes and affections; and the regulation of the purse is, in its essence, regulation of the imagination of the heart." Men and women must be taught *not* to want useless, wasteful things—the very things that multiplied exponentially every week in city shops and department stores. Proper consumption required an education of desire.[17]

This was not merely a moral matter. Ruskin also worried about the limits of natural resources. One of his interlocutors disturbed him by claiming that "The wealth of this world" could be "'practically' regarded as infinitely great." Ruskin was vehemently opposed to this idea. He retorted that "[t]he healthy food-giving land, so far from being infinite, is, in fine quality, limited to narrow belts of the globe." Nor were "animals" infinite, or "minerals ... iron, coal, [or] marble." They existed in "large quantities," but this was no assurance of an endless supply: "[S]o much for the infinitude of wealth." He also ridiculed the waste plaguing modern consumption. He took issue with the *Times* for claiming that "'by every kind of measure, and on every principle of calculation, the growth of our prosperity is established,' because we drink twice as much beer, and smoke three times as many pipes, as we used to." He noted sharply that neither the man who does so, nor his wife and children, will be "materially better off for it."[18] Further, Ruskin predicted that the steam and coal economy that made Victorian industry possible would eventually grind to a halt. He was aware of the famous forecast made by the political economist William Stanley Jevons in 1865 that Britain would run out of coal within a century or so. Ruskin specifically mentioned the likelihood of coal exhaustion in *The Queen of the Air* (1871). While the finite stocks of coal were quickly used up, the renewable resources of wind power and human muscle were neglected. It was an "insane ... waste" to favor coal in these circumstances. In fact, "We waste our coal, and spoil our humanity at one and the same instant. Therefore, wherever there is an idle arm, always save coal with it, and the stores of England will last all the longer." Ironically, for Ruskin the idea of exhaustion promised a bittersweet outcome. Once coal had finally and truly run out, its absence would make it easier for people to recognize the true aim of political economy. But in the meantime, coal was something to use with caution rather than squander.[19]

"THERE IS NO WEALTH BUT LIFE," Ruskin declared in *Unto This Last*, the book he thought posterity would remember "when men have forgotten every word of my *Modern Painters*." It was here that Ruskin set out most clearly his notion of the ethics of consumption. He attacked the idea that demand was something relative and therefore indefinite: "Economists usually speak as if there were no good in consumption absolute." By this, Ruskin meant that liberal thinkers failed to consider the moral aim of the process. "So far from this being so, consumption absolute is the end, crown, and perfection of production." From this followed that "wise consumption" was "a far more difficult art than wise production." The "vital question" to ask was "never 'how much do they make?' but 'to what purpose do they spend?'" It mattered greatly whether you purchased items merely to show others how much you spent or whether you *purposefully* bought well-made things. This problem of buying quality goods was connected in turn with reducing consumption. For poor people, this meant living within one's means and being satisfied with one's station. What the country needed above all, Ruskin argued, were people who "have resolved to seek—not greater wealth—but simpler pleasures... making the first of possessions, self-possession." Such an "art of living," Ruskin thought, was best learned at home, not through formal education. This emphasis on household instruction went hand in hand with Ruskin's general rejection of social egalitarianism: "I am not... countenancing one whit, the common socialist idea of division of property." He insisted that the moral example of the elite would do far more good: "The rich man does not keep back meat from the poor by retaining his riches; but by basely using them... The socialist, seeing a strong man oppress a weak one, cries out: 'Break the strong man's arms' but I say, 'Teach him to use them to better purpose.'" Ruskin's ideal world would still have consumers; they would just be carefully schooled—by responsible elites—in the merits of wise spending and self-restraint.[20]

Above all, then, Ruskin was interested in the problem of how to use commodities in the service of life. How much should a person consume? What things were most needful? He hoped to channel consumer appetites in new directions by inspiring an ethos of material simplicity, compassion, and the protection of natural beauty. However, he was less than certain that there would be any natural beauty left to protect in the grim future.

THE STORM CLOUD OF THE NINETEENTH CENTURY

Ruskin's high hopes for a new kind of moral economy were wedded to apocalyptic anxieties. He announced that the atmosphere of Europe was undergoing a dramatic and unprecedented perturbation in two public lectures on "The Storm Cloud of the Nineteenth Century" at the London Institution on February 4 and 11, 1884. These lectures were well attended and drew reporters from the *Pall Mall Gazette*, the *Art Journal*, and the *Daily News*.[21] To the incredulous audience, Ruskin declared the beginning of a new epoch in natural history, the era of the Storm Cloud:

> This wind is the plague-wind of the eighth decade of years in the nineteenth century; a period which will assuredly be recognized in future meteorological history as one of phenomena hitherto unrecorded in the courses of nature, and characterized pre-eminently by the almost ceaseless action of this calamitous wind.

This new force in nature had "a malignant *quality*" alternating between "drenching rain," "dry rage," "ruinous blasts," "bitterest chills," and "venomous blights." The wind was trembling and intermittent, changing rapidly from state to state. It came from all quarters of the compass, but seemed to favor the southwest. The phenomenon was not confined to any single district of Britain or Western Europe. On his journeys, Ruskin had noted its range "from the North of England to Sicily," where it blew "more or less during the whole of the year, except the early autumn." Ruskin added that the new plague-cloud was "*always* dirty, and *never blue under any conditions*." It had the power to eclipse sunlight "in an instant."[22]

Throughout his lecture, Ruskin struggled to describe what he also struggled to *see*. In an attempt to chart the main features of the phenomenon, he showed drawings and paintings made over the course of his life, the first one from 1845. Of all the features of the Storm Cloud, the tremulous character of the wind seemed to occupy Ruskin's imagination the most. This eerie phenomenon made tree leaves shudder, "as if they were all aspens . . . with a peculiar fitfulness" that gave them "an expression of anger as well as fear and distress." It was perhaps the subtle quality of the tremors that so greatly unnerved him, as though he felt the burden of being the only one able to assess what was going on, little by little, over so many years. He repeated verbatim in the London lecture an entry from his Brantwood diary made on July 4, 1875:

> [A]n hour ago, the leaves at my window first shook slightly. They are now trembling *continuously*, as those of all the trees, under a gradually rising wind, of which the tremulous action scarcely permits the direction to be defined,—but which falls and returns in fits of varying force, like those which precede a thunderstorm—never wholly ceasing.[23]

The discovery of the Storm Cloud was the product of Ruskin's lifelong habit of drawing skies and obsessively taking notes on atmospheric phenomena. He had made his reputation as an art critic by praising Turner's cloudscapes, where smoke and smog played a central role. Ruskin's obsession with atmospheric change was partly triggered by the dramatic increase in air pollution over Britain. At times, he explicitly linked the degradation of the air to industrial production, describing it as a "lurid, yet not sublimely lurid, smoke-cloud" and a "dense manufacturing mist," born of the "Manchester devil's darkness." The London lectures followed directly the eruption of the Krakatoa volcano in May 1883, which killed tens of thousands and spread so much ash in the atmosphere that the average global temperature sank the following year. All the same, Ruskin denied that the Storm Cloud was *purely* a physical form of pollution, whether of industrial or volcanic origin. At times, he seemed to favor supernatural over physical causes:

> It looks partly as if it were made of poisonous smoke; very possibly it may be: there are at least two hundred furnace chimneys in a square of two miles on every side of me. But mere smoke would not blow to and fro in that wild way. It looks more to me as if it were made of dead men's souls.

Exactly this mingling of causes is revealing. The material and the metaphysical were intertwined in Ruskin's mind. Speaking in the voice of an Old Testament prophet, he traced a straight line from the human heart to the natural world: "Blanched sun,—blighted grass,—blinded man." The Storm Cloud was the product of moral corruption, which Ruskin associated with the industrial economy and modern science. Indeed, the very scale of the phenomenon—the discovery of an all-pervasive change in the atmosphere—indicated a crisis that spanned the whole range of the physical world and of human civilization.[24]

This great pollution of the atmosphere, Ruskin explained, had been caused by Blasphemy: "[F]or the last twenty years, England, and all foreign nations, either tempting her, or following her, have blasphemed the name of God deliberately and openly." On the face of it, this was a bizarre diagnosis. How could an act of blasphemy—even at the collective level—cause an alteration of atmosphere? Yet, as so often with Ruskin, he was using a familiar concept in his own peculiar way, expanding the term to mean something more than mere irreverence of the tongue. Rather, "Blasphemy" was the tendency of the "vulgar scientific mind" to degrade the "good works and purposes of Nature" and reduce them to the most ignoble and ugly particulars. The public had come to see the world as nothing more than a space for competition and exploitation—"every man doing as much injustice to his brother as it is in his power to do." At the end of the Storm Cloud lecture, Ruskin made his targets clear: free-trade policy and rampant capitalism had destroyed the agrarian basis of the British economy and left Britain dependent on American food imports. The country could no longer feed itself but had to purchase food from other countries in exchange for manufactured goods. At the root of this national disgrace was the apology for "usury and swindling" in liberal economists such as Adam Smith and John Stuart Mill. Their defense of market exchange and free trade had distorted and degraded the dignity of human life. The Storm Cloud of the Nineteenth Century was a physical and spiritual manifestation of the corruption of human desire: "Whether you can bring the sun back or not, you can assuredly bring back your own cheerfulness, and your own honesty . . . you can cease from the insolence of your own lips, and the troubling of your own passions." Blasphemy, meaning capitalism and materialism, was a world-destroying force with both physical and spiritual causes and consequences. Against the worldview of modern science and political economy, Ruskin invoked the wisdom of Solomon: "Better is a dry morsel, and quietness therewith, than a house full of good eating, with strife." By rejecting the abundance and competition of industrial society, people would rediscover the peace of modest living.[25]

It is tempting as well to attribute the Storm Cloud, if not to Ruskin's religious upbringing, then to his recurring bouts of mental illness. His increasing susceptibility to mental breakdown—fevers and days of catatonic bedrest mingled with hallucinations—may have contributed to his interpretation of the atmospheric effects he perceived. Perhaps the illness made him more keenly attuned to the slightest alterations, the most miniscule vibrations in the air. He dated his discovery of the Storm Cloud to the year before he moved to Brantwood, precisely the time of his earliest breakdown: "I first noticed the definite character of this wind, and of the clouds it brings with it, in the year 1871."[26]

Yet Ruskin had in fact begun to worry about environmental degradation long before he suffered any mental or physical breakdown. His fears for the health of society and the stability of the natural order grew steadily over the course of many years. He had expressed a sense of crisis as early as 1860, in the final, fifth volume of *Modern Painters*, a full decade before his discovery of the Storm Cloud. Already at that point, he had begun to associate modern industrial society with the threat of global destruction. The thought occurred to him as he compared the cities of two great painters—Giorgione's Venice of the Renaissance and J. M. W. Turner's industrial London. The beauty and splendor of Giorgione's city contrasted starkly with Turner's world of "dinginess, smoke, soot, dust, and dusty texture . . . dunghills, straw-yards, and all the soilings and stains of every common labor." Both men were intimately familiar with suffering and death, but in Turner's time the scale of destruction had fundamentally changed, thanks to the "work of the axe, and the sword, and the famine." For Ruskin, the conflict of the Napoleonic Wars and the rise of industrial society had turned the whole world into a theater of death:

> Full shone now its awful globe, one pallid charnel house,—a ball strewn bright with human ashes . . . all blinding white with death from pole to pole,—death, not of myriads of poor bodies only, but of will, and mercy, and conscience.[27]

In the fifth letter of *Fors Clavigera*, from 1871, Ruskin warned that the forces of industrial production were threatening to unmake the fabric of the natural world. Modern consumers now wielded Promethean power over the atmosphere, oceans, and land: "You can vitiate the air by your manner of life, and of death, to any extent. You might easily vitiate it so as to bring such a pestilence on the globe as would end all of you." This new force also extended to modern methods of conflicts. Ruskin described the Franco-Prussian War as a perverse chemical experiment, spoiling the earth "with corpses, and animal and vegetable ruin," which deteriorated in turn into "noxious gas." He saw the same forces destroying the land in peacetime Britain as well—the "horrible nests" that stood in for reasonable towns were "little more than laboratories for the distillation into heaven of venomous smokes and smells, mixed with effluvia from decaying animal matter." Such vitriol sprang from more than just a gut feeling of general decay. Ruskin was elaborating here on an image of environmental limits drawn from Mill's *Principles of Political Economy* (1848). In the chapter on the stationary state, Mill envisioned a planet so densely populated and exploited that "every rood" (i.e., a quarter of an acre) was in cultivation. In such a world, all wilderness would be plowed up, every animal species would be either domesticated or exterminated, and every "wild shrub" would be "eradicated as a weed." But where Mill had imagined deterioration in terms of overcrowding, extinction, and the loss of solitude, Ruskin instead saw environmental change as a product of spiritual decline, atmospheric pollution, modern warfare, and anthropogenic climate change.[28]

A DETERIORATION OF CLIMATE

Already in 1863, Ruskin had written to his father from Chamonix that the glaciers near Mont Blanc seemed to be in retreat: "Another summer or two will melt the lower extremity of the Glacier des Bois quite off the rocks." When he visited the same spot eleven years later, he discovered that nearly all of the Glacier des Bois had vanished. There was just a "little white tongue of ice" left in the "blank bed" to show where it had once been. But "the saddest of all," he told his friend Charles Eliot Norton, was the thawing of the snow on Mont Blanc. It seemed to hold steady only along the more level parts, but had disappeared where the gradient was steep. In the "Bionnassay valley it [had] all flowed down and consumed away." This alarming contraction was part of a broader pattern. In 1865 he mourned the disappearance of his favorite landscapes by "violent . . . physical action." The Lac de Chede had been filled up by "landslips" from the Rochers des Fiz. Lake Lucerne was growing ever narrower and might soon be split in two. At the same time, the glaciers "north of the Alps" were steadily diminishing, while the "sheets of snow on their southern slopes" shrank at an ever greater speed. Near Pisa and Venice, the Maremma marshes were expanding, spreading foul air and disease. Writing from Vevay on Lake Geneva, he expressed the same dismay about the state of the "higher Alps." Over half a lifetime of visits to Switzerland, he found that the light of the mountains had grown "umbered and faint" while "languid coils of smoke, belched from worse than volcanic fires," defiled the air. The water of the lakes too was now polluted. The "glacier waves" were "ebbing, and their snows fading."[29]

In his efforts to better understand the processes he was observing, Ruskin began to take a special interest in the deeper structure of these glaciers. From the 1860s onward, he came to be embroiled in a bitter dispute with John Tyndall, Professor of Natural Philosophy at the Royal Institution, about the causes of glacial movement. Ironically, Tyndall actually confirmed Ruskin's concern about the contraction of the Swiss glaciers. At the very beginning of his book *The Forms of Water in Ice and Rivers, Clouds and Glaciers* (1872), Tyndall announced that the ice sheet at Mer de Glace had shrunk considerably in just twelve years. The glacier "exhibited in a striking degree" a great reduction in size. If this trend continued unabated, it would "eventually reduce the Swiss glaciers to the mere specters of their former selves." Yet, as Ruskin complained, Tyndall had nothing more to say about this dramatic change in the book. Whereas Ruskin suspected that the melting of the glaciers was part of a frightening broader pattern of environmental decline, Tyndall merely put forth his observation without any investigation of causes and effects. Hardly by coincidence, Ruskin later selected for his Storm Cloud lectures a passage from Tyndall's book to illustrate the blasphemous attitude of modern science. He himself felt the changes were much more ominous, declaring in 1873 that "one third, at least, in the depth of all the ice of the Alps [had] been lost in the last twenty years." This was in fact a "change in climate without any parallel in authentic history." The melting of the glaciers, he surmised, had grave implications for the "water supply and atmospheric conditions of central Europe." He believed it was "by far" "the most important phenomenon" for men of science to study.[30] Soon both disturbing anomalies—the Storm Cloud and glacial melting—began to converge in Ruskin's mind. He ventured in November 1875 that "the black plague cloud" was "probably" linked "with the diminution of snow on the alps." The regional phenomenon of glacial retreat was an expression of the global degradation caused by the Storm Cloud. This notion of "teleconnection" must have taken root firmly in his mind, for two months later he announced that the spiritual depravity of the age—its "Blasphemy"—was producing unprecedented psychological and environmental effects. God was punishing humanity with "widely infectious insanities" as well as "grievous changes and deterioration of climate."[31]

Again, Ruskin's tendency to resort to theological language did not eclipse his interest in physical explanations. Two years later, in February 1878, he discussed the question of climate change at some length with the correspondent Henry Willett, a Brighton businessman and fossil collector. The occasion was the translation of a book about Mont Blanc by the architect Eugène Viollet-le-Duc. The French critic proposed that there might be a link between the melting of the glaciers and the retreat of forests from the higher slopes. This hypothesis provoked in the pages of *Fors Clavigera* a wider discussion of climate change in Britain and on the Continent. Ruskin asserted that "the destruction of the woods on the mountains" had "increased rainfall in the plains" of Europe. He also suggested that the "coldness of the summers" had made it more difficult for clouds to ascend so high as to produce snow "on the . . . summits." Willett in turn analyzed the causes behind the growing frequency of floods in the Oxford region and the English uplands. He too blamed deforestation but added several other factors, including the drainage of wetlands, the paving of roads, and the straightening of rivers. Thus, a full six years before the momentous Storm Cloud lectures, Ruskin and Willett agreed about the underlying cause: human attempts to reshape the landscape were now reaping a harvest of unintended consequences.[32]

A visit to Mont Blanc in 1882 further confirmed the scale and depth of the calamity. Ruskin wrote in despair to his old friend Charles Eliot Norton, now Professor of Art History at Harvard, that the mountain peak they once knew had vanished completely: "All the snows are wasted, the lower rocks are bare, the luxuriance of light, the plenitude of Power, the Eternity of Being, are all gone from it." Snow and air were changing together: "as the glaciers, so the sun that we knew is gone! The days of this year have passed in one drift of soot-cloud, mixed with blighting air." Ruskin's secretary, William Gershom Collingwood, took careful notes of his master's reactions in his journal and sketched the scene with its darkened skies. All the way up the mountain, the windows of their carriage were shut to keep out the "plague-wind." Ruskin's fears lasted throughout his life. Another witness heard a similar story when he chanced upon Ruskin at Sallenches in 1888. That year the peasants of Switzerland had suffered greatly from the wet season. Indeed, Ruskin remarked, "the whole climate of Europe" was "growing damper." The snows of Mont Blanc had thinned a great deal, making the "top quite bare."[33] In *the Art of England* course, given in 1884, just a few months after the scandalous London lectures, Ruskin defined the Storm Cloud as a general "deterioration of climate." He connected natural destruction with the decay of artistic sensibility. The change in climate

FIGURE 1.4 The Smoke-Cloud over Lake Geneva, 1882, by W. G. Collingwood: "Lake of Geneva and Dent D'Oches Under the Smoke-Cloud, From St. Cergues, September 1882." *Ruskin Relics.*

might "paralyze" the sensitive natures of most young artists. In a degraded natural world, they might not even develop the skills necessary to apprehend beauty and sublimity in the world.[34]

The melting of the alpine glaciers was no figment of Ruskin's imagination, although the phenomenon that so troubled him was not caused, as he believed, by the Blasphemy of modern progress. Modern glaciology has measured the uneven retreat of the Swiss ice sheets after the middle of the nineteenth century. Ruskin could not have understood it at the time, but his fascination with the sublime world of the high peaks made him an unwitting eyewitness to the end of the Little Ice Age. Over the early modern period between 1500 and 1800, temperatures across the world had decreased by one degree centigrade. This long chill fed the glaciers of the Alps, permitting them to expand in length as well as size. But in the past century and a half, warming temperatures have reversed the process. Ruskin's keen eye, strong memory, and numerous visits to the Alps between 1835 and 1888 gave him a privileged vantage point to observe this transition. Although he mistook a pattern of variability for anthropogenic climate change, Ruskin correctly identified human influence, including coal burning and deforestation, as a threat to the stability of the climate over the long term. Indeed, already in Ruskin's own lifetime, greenhouse gases had begun their steady climb above the bounds of natural variability. The prophecy of the Storm Cloud anticipated the growing anxiety about anthropogenic climate change in the late twentieth century. Ruskin may have formed his views partly in error, but he was not wrong in his intuition that industrial civilization could for the first time alter the fabric of the natural world on a planetary scale. This basic recognition about the global impact of consumer society has turned out to be eerily correct.[35]

BRANTWOOD

Ruskin tried to escape the storm cloud by moving to the north of England in 1872 (he had purchased Brantwood the year before). But in what should have been the clear skies of England's northern countryside, he began to notice sooty clouds. One of the first additions made to his new home was the construction of a turret on the southwest corner of the house. Through the elegant latticed diamond-shaped windowpanes, Ruskin's gaze could sweep north, west, and south across the lake and the fells of the Old Man as he anxiously charted fluctuations in the atmosphere. When the wind blew from the furnaces in nearby Barrow on the coast, it was obvious that the Storm Cloud had found its way into the heart of the Lake District. Even at Brantwood, he failed to escape the effects of modern society. The prevailing winds from the southwest brought smoke from the new furnaces on the coast. Such disturbing atmospheric phenomena haunted him. These sights, filtered by Ruskin's uncommonly sensitive powers of perception and intellectual acumen, led him to the verge of a mental breakdown. The whole fabric of the natural world seemed to be coming apart. Ruskin wrote in his diary in 1877, "Monday, half sleepless night again—and entirely disgusting dream." "[M]en using flesh and bones, hands of children especially, for fuel—being out of wood and coals."[36]

Nevertheless, the new home in the north provided him with many distractions from these growing fears and a place to mull over solutions. In 1871, just before he moved to Brantwood, Ruskin had conceived of the Guild of St. George, an association devoted to bringing common workers closer to nature and teaching them the benefits of self-sufficient gardening. He was eager to think more deeply about the nature and aims of the movement. Having come to Coniston still a bachelor and mourning his father and mother, Ruskin found solace and a renewed purpose in physical work on the Brantwood grounds. It was not unlike the experience intended for the guild's workers, but because these were his grounds, and they were relatively large, there was much flexibility in his approach to getting his own hands dirty. The whole of the Brantwood grounds reflects the mental and physical energy of a working garden, a site of active moral and botanical experimentation. This was always Ruskin's intent, though he might have bristled at the word *experiment*. In his mind, he was simply returning to an older mode of land use, free from the depravations of the industrial age. To this end, there were several different gardens established according to varying themes. The Professor's Garden most overtly addressed social issues. Ruskin used it to gain greater insight as to what might be expected of a worker's allotment, to learn how the hoe and spade and one's own small plot of land might improve the lives of workers. He was adamant that it should reflect a concern for spiritual as well as physical health. Vegetables and herbs, useful for the body, grew alongside flowering vines, a comfort for the weary mind.[37]

Body and mind could also be replenished in other ways. Ruskin's diary entries and letters contain numerous references to chopping wood. The activity seems to have served as a spiritual exercise of sorts, focusing the mind on the hard labor and thrift of producing local fuel for the house. A girl in the neighborhood whom Ruskin called "my little wood-woman" came "every day" to collect the sticks he chopped and "fill the log basket—by the study fire." Brantwood was always drafty, damp, and cold, and Ruskin probably did not expect to gather enough wood to warm up his whole sprawling house. But woodcutting held a deeper meaning than the mere heating of a house. He complained bitterly about the high price of coal and scoffed at newspapers that touted coal consumption as a sign of the increasing affluence of the general population. By their reckoning, the more coal was sold, the wealthier the nation must be. Though Ruskin wondered: how could coal be counted as a form of wealth if the poor could ill afford it? Chopping one's own wood was thus an act of self-reliance, a struggle for independence on the part of the common man, and a commentary on the specious optimism of liberal economists. He also believed wood chopping could be a means of avoiding pollution. In 1877 he told the readers of *Fors Clavigera* that "I have been very busy clearing my wood, and chopping up its rotten sticks . . . I am highly satisfied with the material results of this amusement; and shall be able to keep the smoke from my chimneys this winter of purer blue than usual, at less cost."[38] At the time, wood smoke produced far fewer pollutants than coal smoke. Further, if properly managed, wood could be a renewable resource. Ruskin scathingly dismissed an article in the *Edinburgh Review* (1875) on the wonders of coal as fuel. Mineral stock was not only grossly polluting but a pathetically limited resource in the long term: "while the life of the coal-field has been taken at 150 years, that of the forest, if rightly cared for, will endure as long as that of the human family."[39]

Even today, a tour of Brantwood conveys much of Ruskin's original design. A great number of experiments and trials are underway outdoors. The gardens are kept up to perfection, but not in the manner of evenly trimmed hedgerows and carefully weeded flower beds. Ruskin placed great emphasis on the idea of working with Nature, rather than against it, and this ethos remains in effect. What could be done in the garden other than weed, trim, shape, or force green things into a pleasing pattern? One could think about the land; one could approach nature not by means of slides and microscope, but with one's own eyes and hands. It is often repeated that Coniston is and always has been a working village. Ruskin was enamored with the idea of a place where wild nature thrived with the aid of human hands. His Moorland Garden was designed to uncover new ways of growing food on land not normally considered fertile. It was to be a small working model of the gardens he hoped to set up all over the country through the Guild of St. George. He tried to grow wheat in a hillside clearing above his house to prove that marginal lands might have hidden uses. He had his young students set their spades to building an irrigation system on the high plateau as well. Though the project failed to grow wheat, the enterprise nevertheless succeeded in putting hands to work—which was good for the soul, he believed—and in paving the way for future trials. Later he grew cranberries, fruit trees, and wildflowers. Ruskin made some concessions to more conventional modes of gardening, against his better judgment, largely because of his cousin Joan, who was to inherit the estate. He included greenhouses, for instance, though they were at first unheated. To counter accusations of frivolity, he ensured that they produced more than luxuries for the Brantwood table; they were used to grow grapes for the local people when they fell ill. Similarly, the ice cave on the slope above the house provided ice to soothe patients suffering from fevers.[40]

At the base of the hillside on a thin strip of more-or-less level ground above the lake, the modern visitor finds the parking lot of the Brantwood museum. It appears somewhat incongruous in a world otherwise bursting with plants, trees, and shrubs. This is where the greenhouses and vegetable gardens sat in Ruskin's time. Brantwood's present-day survival depends on a relatively large space for visitor parking precisely in the spot where the land once provided Ruskin's household with a measure of self-sufficiency.

FIGURE 1.5 Ruskin's Moorland Garden near Lawson Park, where Agnes Stalker's family lived. *Ruskin Relics*.

But then, Ruskin was given to a few contradictions in his own life. He maintained his own gardens with a light touch, refraining from imposing exotic flowers on the rugged landscape, nurturing what has grown there for centuries. His gardens embody this ideal in countless ways. But one exception was the coppice woods on the eastern slopes of the lake. These lands had been used by humans for centuries to produce wood for charcoal, poles for fence building or other Lakeland crafts, and even for early iron smelting. Ruskin loved his woods and went out nearly every day with a small axe to tend them. But his aim was not so much to revive ancient sustainable practices as it was to grow the trees to match the ones he admired in the paintings of Botticelli. Whereas proper coppicing requires that the main tree be cut down periodically—so that new shoots can sprout from the base—Ruskin chose to take out the younger shafts and leave the older, larger trunks intact. They grew tall and thick and imposing, reminiscent of an Italian hillside. This resulted in more heavily shaded areas that would in time have choked out most of the ancient undergrowth. One can sense here uncomfortable tensions between Ruskin's love for nature and his commitment to aesthetic ideals. Yet, happily, sometimes worthy ideals are better realized by one's disciples. A plan is now underway to resume the ancient industry of coppicing to ensure the health of the surrounding woods. There are lime trees above Brantwood that are 500 years old. They might live longer still, now that they are being properly managed and thinned.[41]

Rising above the parking lot is a strange addition to the landscape, the "Zig-Zaggy Garden." The idea for this curious switchback path draws on some preliminary sketches made by Ruskin, possibly inspired by Dante's account of the ascent through Purgatory. It is the work of Brantwood's latest master gardener, Sally Beamish. A sign informs the visitor that the design presents "an allegory of the Seven Deadly Sins—Pride, Envy, Anger, Sloth, Avarice, Gluttony and Lust." Starting at the base of the hillside and following a zigzag path upward, one first encounters elaborate terraced planting beds with bunches of black ornamental grasses. Farther up are gloomy arrays of dark blue chips of slate dotted with meager green grasses. These are curiously bordered with tufts of matted sheep's wool. The bed recalls both the slate quarry across the lake and the sheep in the fields, alluding to two occupations that have long coexisted in the Lake District.

This portion of Brantwood's grounds is both striking and unsettling. Despite tiny blooms of rugged sedum clinging to the rocks and the promise of flowers nearer the top of the path, the garden's lower portion hints at blight. It is almost as though soot has fallen from the sky to coat the fine tendrils and spiky shoots. The garden in fact calls to mind the industrial ruins found all around the Lake District. Is this a healthy garden stunted by the Storm Cloud, or a wasteland that is slowly coming to life again? In any case, Ruskin probably would have applauded the attempt to provoke complacent observers to reflect on the fragility of the natural world as they wander up to Ruskin's garden paradise.

As with every utopian critique of society, there was a dark side to Ruskin's project. His vision of small-scale rural simplicity and virtue was driven in good part by a powerful loathing of modern cities and technology. He treated Darwinian science with distaste and skepticism, seeing in the theory of evolution merely a reflection of the godless greed and callous competition he associated with unrestrained capitalism. Further, his theory of consumer ethics reinforced his alienation from the world of contemporary politics, just as his move to the Lake District had taken him into a kind of inner exile, far away from London and Oxford. He had no patience with liberal or conservative politics of the day, rejecting Disraeli and Gladstone alike. Although many of his ideas about social welfare (a term that he actually coined in English) would have a deep influence on radical politics and the construction of the Labour welfare state in twentieth-century Britain, Ruskin himself looked backward in time to Xenophon's political economy, Plato's *Republic*, King Solomon's Israel, and the guilds of the Renaissance for an ideal of good government. Though he criticized the consumer habits of the middle class, he accepted that social inequality was fundamentally natural and even benign. In other words, Ruskin was a Tory radical; that is, a conservative who treated urban bourgeois life with contempt but would not lift a finger in favor of social revolution.[42]

Moreover, it is important to acknowledge Ruskin's difficulties when it came to living by his own ideals. He was a world-renowned Oxford don whose presence was desired all over England and on the Continent. He traveled more than most people, often by train, even though he lamented the destruction wrought by railways on the landscape. He lived a life of privilege thanks to his father's wealth, yet as famous and wealthy as he was, he remained a lonely bachelor, enduring gossip about his unconsummated, annulled marriage to Effie Gray. Matters were not helped by his stormy and ill-fated infatuation with the young, iron-willed Rose La Touche, a girl almost thirty years his junior. When Ruskin first saw the ten-year-old girl in 1858, he was enchanted. In 1866 he shocked everyone when he asked Rose, then eighteen, to marry him. It was the start of a torturous, constantly deferred attachment that ended only when Rose died in May 1875. She had apparently succumbed to a combination of mental and physical illnesses— likely exacerbated by her famous suitor's relentless attentions. Unlike Xenophon's ideal farmer, then, Ruskin became a hopeless bachelor, a man without a true woman of the house. His bouts of madness, which began during his strained courtship with Rose and continued after her death, only heightened his unusual isolation.[43]

As his illness worsened, he became more and more housebound and sometimes fell completely mute. The circumference of his life shrank to the area around Brantwood where he was led on gentle walks by his cousin or visitors. The Master was now silent. Yet the experiment continued. His words were not lost on his followers. Many of them acknowledged the merits of his argument in deed rather than word, digging into their gardens, reviving ancient handicrafts, cultivating the art of living, and extending the boundaries of imagination itself, all setting Ruskin's theories in motion with a quiet but unremitting passion.

CHAPTER TWO

Selling Sufficiency

On one of Ruskin's trips to Italy, he observed how glass beads were made from long rods by Venetian workers. The rods were "chopped up into fragments of the size of beads" and then "rounded in the furnace." Such monotonous work had a disturbing effect on the workmen. Their hands vibrated "with a perpetual and exquisitely timed palsy." This was mindless labor without "the smallest occasion for the use of any single human faculty." Beads dropped like "hail" from their "vibrating" fingers. In *The Stones of Venice*, Ruskin contrasted this drudgery with the thoughtless indifference of the consumer. Ignorant women without taste or knowledge were the drivers of the glass bead industry. Their whims kept these men employed in a form of labor Ruskin held to be worse than plantation slavery. Social and geographic distance allowed consumers to turn a blind eye to the bleak reality of the manufacturing process just as they did with the American plantations. But what would it be like to consume objects with a full understanding of the conditions under which they were manufactured? What social order would arise from this kind of knowledge? In Ruskin's image of the glass beads, ethical consumption was not simply an abstract recognition but also a form of sympathy that reached into the realms of workshops and the bodily experience of the laborers. His words shamed the reader into seeing the hidden processes behind the objects of consumption. Ruskin made the glass beads speak, revealing the nightmarish origin of these innocent-looking baubles.[1]

Over the years in Coniston, Ruskin filled Brantwood with exquisite paintings, hundreds of books, and elegant furniture. He felt that things made well nourished the soul and preserved the beauty of the landscape. Mass-produced commodities, by contrast, poisoned the mind as well as nature. In *The Stones of Venice*, Ruskin introduced three rules to guide consumption: "Never encourage the manufacture of any article not absolutely necessary, in the production of which Invention has no share." He also rejected the appreciation of an "exact finish for its own sake." Standardization and precision had no value in themselves; the quirks and small ingenuities of handcrafted goods were far preferred. Finally, he ruled out "imitation or copying of any kind," except when it might preserve "great works." These rules should also guide manufacturing. Extreme symmetry and standardization were marks of slavery, degrading the worker by making him a mere machine. Instead, labor should combine the work of the hand and mind. This meant an all-too-human possibility of failure and flaws, but also the freedom to create truly "noble" things. Ruskin was of course not alone at this time in calling attention to the dehumanizing routine inherent in the modern factory system. But his vision differed from the revolutionary aspirations of Marx and the socialists. Ruskin looked to the past for his model of good work. *The Stones of Venice* spurned both the "present English" and the ancient Greek mode of production in favor of the "Gothic" style of medieval artisans. As an artistic manifesto it proved brilliantly provocative, stimulating the work of William Morris and the Pre-Raphaelite painters. But as an economic and moral vision, it left Ruskin's followers with a strange and difficult proposition. Was Ruskin's plan in fact a plea for a return to the Middle Ages? If ethical consumption required the revival of handicrafts, how and where could this be done?[2]

THE GUILD OF ST. GEORGE

The movement to revive traditional crafts and industries in the Lakes region between 1880 and 1920 began with Ruskin's series of letters to "the Workmen and Labourers of Great Britain," known as *Fors Clavigera*, the first of which was published in 1871. These long and complex letters, bursting with literary allusions and flights of philosophical fancy, did not coddle the working-class readers but treated them as equals. If the form of *Fors Clavigera* was unapologetically intellectual, the social goal was practical. The key question was how to help the poor and downtrodden of Britain find new material security and independence. There was more than an echo here of Ruskin's other project to introduce Xenophon to "British peasants." They needed first to be shown how to help themselves, and second to be induced to *desire* to help themselves. Ruskin promoted no easy handout of cash, but rather a way of instilling a culture of sufficiency among the vulnerable and anyone who might be tempted to steal or swindle others. A properly educated worker would know how to provide for himself in part because he would have plenty of resources at hand, but also because he would understand the importance of providing himself with just the right amount of useful things.

Here was the seed of a social movement. Ruskin began to develop the Guild of St. George, an organization designed to inspire an ethos of sufficient living among workers in a planned community outside the industrial town of Sheffield in the Midlands. Ruskin was directly involved in the practical business of the guild. He made the first donation himself, amounting to £7,000. He helped procure property that had been abandoned or deemed barren through donations from his readers and followers. All the land was to be turned over to working men to see how much they could reap from it. In its way, the idea was not unlike community gardening plots seen today. By learning how to harvest food from the land, Ruskin hoped the members of the guild might gain a new respect for nature. With their proposed schools, libraries, and art galleries, these institutions would be aimed at reforming the character and needs of the guild members, teaching them to be satisfied with the diet, education, and art of the community.[3]

The Guild of St. George benefited from a small number of early donors—in terms of both land and money—and was soon granted a limited liability license by the Board of Trade in 1878. By 1882, however, the dream of acquiring a vast acreage on which to develop new communities had to be abandoned. The founders had failed to find enough land and funds. Instead, Ruskin began accumulating books and cultural artifacts deemed useful for the moral development of working men, women, and children; these he donated to a purpose-built St. George Museum at Walkley (outside Sheffield) as well as to existing schools in various places. The mission of the Guild of St. George developed further to include the support of business ventures, especially those that directly opposed the industrial model that polluted cities and tied laborers to harsh working conditions. Crucially, the participants in Ruskin's experiment had to swear off the use of coal and steam engines. The guild pointedly endorsed the use of renewable fuel, preferring work by hand and machines "operated by natural force of wind and water," and perhaps electricity.[4]

THE LAXEY MILL

Egbert Rydings was a struggling silk weaver from Lancashire. With money short, he and his wife moved from the mainland to her hometown on the Isle of Man. In time their financial affairs improved, but his wife's health suffered. While he nursed her through a long terminal illness, he began to read Ruskin's works. They struck a chord. In 1875 he began a correspondence with Ruskin, and in time he proposed to start up a woolen cloths industry based on principles derived from Ruskin's political economy. The Guild of St. George eagerly offered assistance. Laxey had once had a thriving spinning and weaving industry, until manufactured and imported fabrics changed the face of cloth production. By the 1860s most workers had been lured into mining jobs instead, leaving a scant number of hand loom weavers and spinners in the area. A decade later, Egbert Rydings noticed the many women who could no longer earn an income at home as spinners. He thought it imperative to find a kind of work for them that revived traditional handicrafts and allowed them to balance industry with home life. The masters of the Guild of St. George agreed, and sent him a check for £25 "to support the spinning and cloth industry of the Isle of Man." But Rydings quickly decided that manual labor simply could not produce cloths of a quality comparable to new imported cloths; he felt that an entirely home-based cottage industry would never succeed. Nevertheless, a suitable compromise was soon found. Water-powered machinery could produce a finer cloth in large volumes without the use of coal and steam. With the acquisition of an old corn mill on the Laxey River, his business venture soon took off, guided by Ruskin's principles. Rydings aimed to ensure that "all materials used in the manufacture would be of the best and purest quality," that all products were made "'as perfect as fingers can make them,'" that consumers could buy directly from the mill without the heavy markups of retailers, and that nothing could be bought on credit. This hard work of making cloth involved many difficult steps; some, though not all, were friendly to the environment and the workers.[5]

First, the items had to be made from 100 percent wool. They could not use "recycled cotton or linen rags," which manufacturers often employed to create cheap, inferior cloth. Women then sorted the wool fleeces and removed any dirt or grass the animal may have picked up in the fields. The wool was washed in "vats of hot water, set over fires of dry gorse." They used a natural soap made from soapwort or "fats and oils boiled with caustic soda." After rinsing and wringing, some wool was transferred to the dye house. After that, all wool had to be dried, spun, and woven, then washed and dried again. Despite Rydings's high ideals, many aspects of the production process were mechanized in the end. Early spinners carded the wool themselves in preparation for spinning, but again, this was deemed too time consuming and ineffective. Instead, at Laxey, the wool was arranged "on the large rollers of a giant [water-powered] carding engine that moved over hundreds of short metal spikes." Terrible accidents were not unheard of—an especially sad state of affairs since young boys often took on this particular work. Oiling the wool helped it run through the machine, but then "Card Room Fog" arose from the pummeling action of the engine against the wool, and this in turn caused a kind of dermatitis. While Rydings used the word *homespun* to characterize his cloth, in reality his workers made use of a "mechanized spinning mule." Likewise, instead of employing hand loom weavers (and, to be fair, there were not many around), he persuaded investors to lend him £200 to buy two water-powered looms, each of which "could match the output of between four and six hand looms." Large sections of cloth were "finished" at a "fulling mill" and then dried by hanging them, while wet, on various wooden frames—a laborious procedure, as wet wool can become quite heavy. Finally, the dried cloths would be sent to a mending room, where young girls inspected the material and fixed any minor flaws with their small fingers. Despite this fine handwork, many steps in the process seemed a far cry from the kind of cheerful cottage industry Ruskin had in mind when he founded the Guild of St. George.[6]

Rydings always walked a fine line in his pursuit of a profitable yet earth- and labor-friendly business. In addition to some of the health and safety problems already mentioned, wool processing led to the pollution of the lower Laxey River. Not only the scouring and dyeing chemicals, but also arsenic might be rinsed from the wool because sheep were often doused with the poison before being sheared. Everything drained straight into the river. Egbert Rydings himself noted the "liquid slime" pouring out of the mill's wastewater drain hole. But at least he attempted to employ natural, locally sourced dyes wherever possible, even as synthetic, imported alternatives were becoming common in the 1880s. However, this choice was not without complications. Although the use of synthetic dyes might have "led to an increase in bladder cancer amongst the dyers," natural dyes like the red coloring from the madder plant could "cause nasty skin rashes" not only in the dyer, but in the person wearing the finished cloth. Another hazard for laborers included noise pollution caused by the mill itself. The constant din "from flapping pulleys and belts, rotating machines and even the swoosh of the water wheel" posed a risk to workers who spent more than a few minutes there. The moving parts of the water mill, water-powered loom, and other machinery also led to injuries. The handling of chemicals such as "soda ash, caustic soda, sulphuric acid, mechanical oils and lubricants" presented unseen perils to anyone involved. Further, power looms could create "lint and fibre dust" that contributed to respiratory ailments. Sadly, balancing sustainable and labor-friendly manufacturing techniques against ambitious production goals proved a complicated task.[7]

Rydings admitted in 1900 that he was "at least £200 poorer than . . . when [he] commenced the business." Yet, in terms of long-term viability, the Laxey venture could be seen as a modest success. Some of Rydings's employees enjoyed nearly two decades of stable, well-paid work, and presumably better working conditions than those employed in the mines. Even though the Guild of St. George stopped supporting the Laxey Woolen Mill when Rydings retired in 1901, the business continued to thrive. Indeed, the Manx woolen industry still manufactures cloths in the twenty-first century, using traditional techniques when possible alongside newer methods of production.[8]

HOMESPUN INDUSTRY IN THE LAKES

Egbert Rydings was not entirely correct in thinking that hand-powered looms and small-scale home-based production could *never* be economically viable. Some thought that cloths could be made in ways that would be even more in keeping with the principles of the Guild of St. George while still remaining profitable. Scholars have shown that the Lakeland arts and crafts industries were the most heavily influenced by Ruskin's principles.[9] Friends and followers who regularly referred to him as "the Master" were in charge of these Lakeland projects. It was not merely Ruskin's ideas, gleaned from his books, that motivated and shaped their efforts; they drew constantly on their daily interactions with the man himself, with his gardens, his household objects (which he sometimes lent them for closer observation), and his unique habits. While industries like Morris & Co. employed its workers in busy studios, Langdale Linen Industry spinners worked from home. The strict rejection of factory work environments meant that these workers produced a smaller quantity of goods, and earned roughly half of what Morris's workers earned. But their products reflected the comforts of home and a greater sense of individuality and responsibility. In this way the Lakes region arts and crafts movement came to resemble a Victorian version of "slow living." Although Ruskin admired the handicrafts of medieval Venice, he had no intention of trying to reintroduce them to a modern city like London or Manchester. Instead, he and his followers focused their attention on rural landscapes, primarily in Cumbria. His critical gaze and appreciation for local materials gave the Lakeland cottage industries a distinctly regional flair.[10]

The first of these projects was launched in Wray, west of Coniston, in 1880. Hardwicke Rawnsley, the vicar of Wray, approached Ruskin to discuss "how to add happiness to the country labourer's lot." The two men agreed that "idle hands should have something found for them to do by other than the Devil": "'We must bring joy, the joy of eye and hand-skill to our cottage homes.'" While Rawnsley heeded Ruskin's words, he was skeptical about the idea of hand spinning and weaving: "[T]he thought of reviving that industry never occurred to one as possible." He believed wood carving might be a craft better suited to the interest of residents of Grasmere, Ambleside, and Wray. He reported that "a lady was engaged to come down from South Kensington to give a course of lessons in the three villages, and our humble home industry in the lake district was set on foot." Instruction included techniques used for metal repoussé as well, a skill "taught with the help of [Edith Rawnsley's] mother's Swiss butler." But when Rawnsley moved to Keswick to take up residence as vicar of St. Kentigern at Crosthwaite, his energies were dispersed in new directions. We will come back to Rawnsley's many projects in Keswick in a later chapter.[11]

Other men and women living in the area started up a number of different cottage industries or handicrafts schools as well. The Quaker furniture maker Arthur Simpson had established a successful carpentry shop by 1885. With the friendship and input of Ruskin's student William Gershom Collingwood, he organized wood-carving classes in Coniston and a regular Arts and Crafts Exhibition. Simpson's furniture business proved quite successful, flourishing until his death in 1922. The firm itself closed for good only after World War Two. In 1904, painters and other fine artists in the region came together to form the Lake Artists' Society under the guidance of Collingwood. There was also a spinning venture in Bowness managed by Annie Garnett, who had met Ruskin and learned how necessary it was "for the artist to go to nature." Her workshop remained open until the 1930s. It is probably a mistake to insist on sharp lines between these different endeavors. True to Ruskin's own example, Collingwood, Rawnsley, and the other leading spirits behind the movement saw no need to separate arts, industry, and scholarship. Collingwood in particular joined together the protection of Lakeland arts and crafts with other kinds of learning, including archaeology, literature, and history. The whole region was alive with activity.[12]

Among these ventures, the Langdale Linen Industry has a special place. Conceived of and promoted by Albert Fleming, the London barrister and follower of Ruskin, this manufacture in Elterwater, Langdale, was the first of the Lakes region handicraft projects to succeed in the 1880s. One scholar goes so far as to say that the whole Lakeland Arts and Crafts movement began with the friendship between Ruskin and Fleming.[13]

THE DISCIPLE AT NEAUM CRAG

"Went over to Little Langdale to lunch with Mr Albert Fleming, an ardent disciple of the Prof.," wrote Walter Druce in 1884. The Ruskin Library in Lancaster recently acquired Druce's diary, which offers a rare glimpse of the daily life of Ruskin as seen through the eyes of a visitor to Coniston. At Brantwood, Druce and his family were treated to splendid meals, walks through the gardens, a hike up the Old Man, and readings of favorite novels with the Professor. A visit to Fleming's home left a more ambiguous impression. In his commitment to the Arts and Crafts, Fleming fit in perfectly with Ruskin's circle; in other ways, he always remained an isolated, strange figure. Highly educated and idealistic, with narrow, finely chiseled features, Fleming was a figure of fierce devotion and considerable eccentricity. Druce wrote in his diary a brief tantalizing account of his excursion to Fleming's home at Neaum Crag. Druce commented on the "beautiful views out of windows over Langdale Pikes—curious house all bedrooms at the bottom . . . originally built by a miner." With surprising nonchalance, Druce added, "He had his coffin in his own bedroom, fond of music . . . Very odd sort of man."[14]

Fleming was just eccentric enough to become enamored with the ancient art of hand spinning and weaving, to learn to spin himself, and to think he could revive this art in the new age of mechanically produced cloths and fibers. The coffin in his bedroom reminded him nightly that his time on earth was short, and that as a religious man he must make the most of it. He kept up a voluminous correspondence with Susanna Beever, many years his senior, detailing his struggle to make Langdale linen a success and sharing with her his love of animals and Lakeland scenery. Beever's letters to him dwelled at length on the small pleasures of Coniston life but also on the solitude and illness of her final years. Fleming too must have felt the threat of decay and decline, especially the vulnerability of traditional livelihoods; he was keenly aware that "all the old trades are dying or dead—bobbin-turning, charcoal-burning, wood-carving, basket-making, hand-spinning and weaving."[15]

This affection for outmoded handicrafts was most peculiar for a man of his class. Fleming could have been reaping the fruits of his profession, living a life of ease in the upper middle class of London. He was a barrister who had inherited his practice from his father and was employed at Gray's Inn. He called it "lucrative, [easy] work" and, he vented, "I hate it." In 1883 Fleming took the leap into a new world. He began to take time off from his practice to enjoy extended sojourns in the north. He wrote to Ruskin about his profound disappointment with modern fashion and wants. Why did so many people feel the urge to sacrifice beauty and dignity in favor of material wealth? The roots of his regard for Ruskin's philosophy went deep. A decade earlier, Fleming had begun to voice some of these dissatisfactions in the pamphlet *The House of Rimmon* (1873): "It has come to me at last, after contentedly worshipping at the . . . shrine of Mammon . . . to ask myself two simple questions. Do I want this money? And how does it better me, having got it?" Fleming wanted to cut through the distractions of status and wealth to discover the basic elements of human well-being: "What are the necessaries of life? Four, and four only: food, clothes, a dwelling-place, and means of education." *The House of Rimmon* ended with an injunction to "resolutely live a noble life."[16]

Fleming's move to the Lake District was an attempt to be close to Ruskin, whom he considered his mentor, and to begin concrete experiments in living a better life. In part, he followed the path of Ruskin's artistic ideals. To understand the nature of good art, Ruskin allowed him to "carry away Jerome—or Melencoly," to study in depth at home. These valuable images by the Renaissance painter and draftsman Albrecht Dürer were held in Ruskin's collection at Brantwood. Fleming also took an interest in fiction writing. Over the years, he published a number of short stories concerned with controversial social issues of the day. But certainly his greatest impact was felt in his endeavors to assist local inhabitants by putting Ruskin's ideas to work for them. In this, he was partly motivated by a desire to safeguard nature against the "rash assault" of modern life. He worked closely with Canon Rawnsley, W. H. Hills, Ruskin, and others to keep railways from defiling the beautiful landscapes. He also imbibed from Ruskin an understanding of the threat of pollution, as the copper mines in the hills sent chemicals and debris down the mountainside into the lakes.[17]

Fleming had long been sensitive to the particular merits of homespun clothing. *The House of Rimmon* sang the praises of handmade textiles from an age before they were "blown together by machines," when they were "neatly fashioned, and made (best of all) with stitching of ready fingers . . . and so, valued and treasured as only home made things ever are." While hand spinning and weaving had been discarded in favor of mechanical techniques on the Isle of Man, Fleming would not give them up so easily. The Langdale Linen Industry offered a practical experiment in the value of handmade goods, the relation of handicrafts to the natural world, and how to market such commodities to the right sort of customers. Fleming's plan was to recruit workers among the poor elderly women in the surrounding neighborhood, whom he thought would be willing to work if they only had the chance. There was a rich tradition of spinning and weaving in the Lakeland region dating back to the late sixteenth century. But with the advent of mechanized looms in the late eighteenth century, the entire industry had collapsed, and the old tradition had been disrupted; grandmothers had given up passing the skills down to their daughters and granddaughters. Ruskin had written on several occasions of the importance of needlecraft to women and humankind in general. He called the needle the "feminine plough" that provided clothing just as the actual plough provided food. He wanted women to master the history and design of each piece of clothing. Speaking of lace making, he observed that the beauty of the craftsmanship depended on the right combination of skilled industry and careful attention to individual details. Fleming believed that this kind of artistic work was best carried out in a household setting, and his proposal to revive domestic linen production was met with great enthusiasm by Ruskin. Soon a plan to set local women spinning again took shape with Ruskin's financial assistance.[18]

Fleming's housekeeper, Marian Twelves, was taught the art of spinning first, but Fleming himself learned to spin soon after. He then commissioned a carpenter to make fifteen more spinning wheels. A workshop space was acquired in Elterwater in Great Langdale Valley and named St. Martin's. The women began to learn the ancient art for the first time in their lives. Whenever anyone "could produce a good thread [Fleming] let her take her wheel home, and [he] supplied her with flax, buying back her thread, when spun." After this first step, Fleming had to figure out how to weave all those new skeins into salable goods. With difficulty, he located and purchased a dilapidated loom all rusted over and eaten away by worms. When it arrived at Elterwater, everyone was utterly perplexed by the unassembled parts.

FIGURE 2.1 Hand spinner from *Songs of the Spindle & Legends of the Loom*.

But Marian Twelves found a solution. As Fleming told the story, "the ingenious lady bethought herself of a certain photograph of Giotto's 'Weaving,'" and determined to put the loom together using Giotto's painting as a model. To Fleming's surprise and delight, the experiment proved a success."[19]

Although the early cloths produced were "wretched—as coarse as canvas, dreadful to touch, and horrible to smell," things did improve. Fleming avoided harsh chemicals in the washing, mangling, and bleaching of these cloths. The only bleaching agents used in production were those "made in heaven's laboratory." "We keep as close to our Homer as we can . . . ," Fleming insisted. Such a reliance on traditional methods of production was admittedly an easier task than the challenge Laxey faced with its wool-based cloths. As Fleming's linens improved in quality, buyers became intrigued by the possibilities of the handmade cloth. A woman "skilled in art needlework" thought their natural, subtle, sun-bleached coloring presented a fine and unusual background.[20]

Fleming fondly quoted Greek and Roman authors whenever he wrote about his project. Spinning and weaving were ancient arts, associated with the fundamental fabric of human society and the thread of fate. Something profound had been lost with the coming of the textile factories. When consumers looked at machine-made cloths, Fleming wanted them not merely to see finer, cheaper materials, but to feel keenly the loss of this venerable tradition. He often repeated Wordsworth's lament for the demise of the spinning wheel. His handicrafts revival promised almost magically to turn back the clock and reinvent the lost art the poet had mourned. But the steady references to Romantic poetry were also a shrewd commercial strategy, a way of tapping into a market associated with taste and tourism. Handmade goods that had seemed outdated and expensive now had literary cachet. He capitalized on every bit of country charm and quaintness by publishing accounts of his enterprise in national newspapers and magazines. Readers across urban Britain would have been struck by the sheer picturesqueness of the whole affair. He used Ruskin's name as well and appealed to the conscience of consumers by describing the remarkable improvements the industry made in the lives of the poor. Ironically, Fleming was by nature reserved and, like many gentlemen of the time with aesthetic aspirations, he found deliberate advertisements rather unseemly. He wrote to W. H. Hills in 1884, shortly after the start of his venture, "I want to popularize [Langdale Linen] this [summer], to put it bluntly to advertize it, but I am rather in the position of the decayed lady who had to cry watercresses in the street, 'but hoped no one would hear her.'" Grasping after buyers in this manner carried the taint of greed and desperation, yet the entire business seemed to hang on reasonable sales figures. Fleming decided to generate publicity by placing articles in the right newspapers and journals with a few illustrative drawings. It was a reluctant strategy of popularization that he would return to again and again. The point was not just to promote and expand enterprise, but also to explain and promulgate the ideas behind it.[21]

Fleming's idealism was informed by a shrewd understanding of commercial realities. The new model for manufacturing and consumption had to appeal at both the social and moral level, inspiring the consumer with the desire to buy clothes made by a satisfied, healthy worker in the country.

> EXCUSE is needless, when with love sincere
> Of occupation, not by fashion led,
> Thou turn'st the wheel that slept with dust o'erspread;
> My nerves from no such murmur shrink,—tho' near,
> Soft as the Dorhawk's to a distant ear,
> When twilight shades bedim the mountain's head.
> She who was feigned to spin our vital thread
> Might smile, O lady! on a task once dear
> To household virtues. Venerable Art,
> Torn from the poor! yet will kind Heaven protect
> Its own, not left without a guiding chart,
> If rulers, trusting with undue respect
> To proud discoveries of Intellect,
> Sanction the pillage of man's ancient heart.
>
> —Wordsworth.

FIGURE 2.2 William Wordsworth's poem lamenting the demise of hand spinning, included in *Songs of the Spindle & Legends of the Loom*.

In *Fors Clavigera*, Ruskin had already linked the use of artificial chemicals to factories and poor labor conditions. He published a letter from a woman whose mother was a Cumberland spinner. She described her mother's freshly woven linen "bleaching on the orchard grass," a vivid memory from her childhood. This woman heaped scorn on modern production methods: *"what factory, with its thousand spindles, and chemical bleaching powders, can send out such linen as that, which lasted three generations?"* Similarly, it seems that the Langdale Linen Industry followed Laxey's lead and used only natural plant dyes, even though few would have felt concern about the use of artificial dyes known as "anilines." By obtaining natural dyes from "lichens, grasses, heather and logwood," Fleming offered a novel sales pitch centered on the regional origin of the product, featuring designs and fabrics associated with local materials.[22]

There were other ways to interest consumers in handicrafts as well. Under Marian Twelves's skillful direction, the Langdale Linen Industry shifted the focus of Lakeland spinning and weaving from the production of basic necessities to artistic novelties. Early handicrafts included linen sheets, but the list was expanded to feature items of all different sizes and values: "curtains, portières, tablecloths, chair-backs . . . and occasionally clothing." A lot of the textiles featured embroidery. Unlike the Laxey industry on the Isle of Man then, Twelves and Fleming were putting out a varied array of hand-decorated household goods. Twelves was especially keen on promoting the individuality and quality of the goods. Many of the designs that workers invented were inspired by natural imagery. Some garnered special attention. There was an "invalid girl" who came up with her own designs; a magazine article mentioned "an old lady over eighty years of age" whose embroidery was "a marvel of skill and delicacy." In time, Marian Twelves herself developed a beautiful and complicated pattern she called "Ruskin Lace." This was likely influenced by Ruskin himself, with whom she corresponded and eventually met. He had been collecting textiles from Sicily and Greece for some time, and Ruskin Lace bears some resemblance to his Greek samples. Thus, Ruskin's ideal lady of good taste now had the opportunity to purchase fine handcrafted lace at home in England. By 1889, the Langdale Linen Industry had increased the number of goods they were selling still more: "pincushions, bedspreads, napkins, doyleys, sideboard cloths, slippers, book covers, children's pinafores, various mats and other items of clothing such as linen dresses, modesty vests and dress collars." Marian Twelves proved to be a tenacious and innovative purveyor of Ruskin-inspired products. In many ways, Twelves enjoyed unusual freedom and professional respect by managing the linen industry. She admired Ruskin's ideas and felt compelled to put them to practical use. However, all was not as idyllic as one would hope.[23]

A SUDDEN DEPARTURE

A skillful and articulate woman, Marian Twelves chafed at the stubborn social hierarchies and gender conventions that governed the Langdale Linen Industry. On the surface, Fleming had the trappings of a progressive and egalitarian employer. *The House of Rimmon* had sharply criticized Victorian mores about gender. He deplored the fact that women were treated as wards and pawns by their fathers and husbands: "You have taken all your women-children, and wrapped them up in cotton wool, and have said to them, 'Dear girls, the objects of life are to look pretty, dress well, get married, and win a position in society.'" This sort of treatment could reduce a capable, intelligent person to a "padded, be-crinolined, chignoned, monstrosity scarcely able to put a pin in straight, much less add up her butcher's bill properly." The Langdale Linen Industry offered women a remarkable opportunity to learn new skills and to consider the complex writings of Ruskin and Wordsworth (they held tea parties to discuss books). Here, women were able to design and create their own works, and Fleming gave Marian Twelves in particular significant managerial powers. She had become much more than his housekeeper. Fleming also showed sympathy for the plight of women in general both in his correspondence with Susanna Beever and in several works of short fiction. Some of his stories explored the unfair treatment of unwed mothers or the hardships faced by poorer girls and women who lacked a responsible male guardian. However, it should also be said that Fleming questioned women's intellectual capabilities at least twice in these stories. So, although he was empathetically supportive on many important points, he appears to have had difficulty attributing to women an equal professional status. Championing their cause in a published pamphlet was one thing; dealing with ambitious and intelligent women in the workplace was another. Twelves was one of the driving forces behind the Langdale Linen Industry from its inception in 1883 onward, yet Fleming was exceedingly reluctant to acknowledge her efforts. As one scholar points out, Fleming's published accounts of his experiences mention her only indirectly as a "kind friend," and only once. In his other writings, he failed to mention her at all. It may be that Marian Twelves, who had much to thank Fleming for, could not break through a deep-seated paternalism when she began to voice her concerns over central issues of class, remuneration, and work hours.[24]

Twelves made several oblique references to a quarrel she had with Fleming at the Langdale Linen Industry between 1883 and 1888. As she told the story, she apparently disapproved of how the organization failed to observe strictly Ruskinian principles. Proceeds from sales were split among the entire cooperative of workers. Fleming said as much: "All money produced by the sale of linen is paid into the bank, and the profits will be divided among the workers at the end of the year." Yet the definition of a "worker" may have been a generous one on occasion, for Twelves was rankled by the fact that "people of independent means" had been allowed to sell their crafts "at an exhibition of village industries." Twelves later claimed that Ruskin's "economic principles" had been violated (although it is not entirely clear how or why). She sought "more independence and reliable support" for her own efforts—perhaps in reaction to the unfair and unnecessary help she felt had been given to those of a higher class.[25]

Twelves read Ruskin's work closely. Precisely how well she was able to discuss it is not known, but in a letter Ruskin expressed delight and even surprise that she had been able to glean something useful from his books. This sort of condescension seems not to have troubled Twelves—perhaps because it came from such an admired figure. After all, she was determined to use her impressive manual skills and creative talents in the service of Ruskin's aesthetic and economic ideals. As a scholar on this period of Lakeland Arts and Crafts explains, "Whatever her formal artistic shortcomings, she had a natural ability to digest and interpret Ruskin's often complex writings, and a keen judgment for good, basic design." Probably Twelves did not mind that Ruskin found her intellectual aspirations wanting, as long as he understood how she translated his ideas into things. But she demanded to be heard by Fleming, one way or another. Twelves's accusations that the Langdale Linen Industry was veering off course are somewhat difficult to piece together. But one clue to her feelings may be found in her complaint that undue emphasis had been laid on the production of great quantities of linen, rather than on the quality of items "accomplished by individual effort." Twelves had a point. It seems that whereas Fleming's main priority was to reconcile ethical production and consumption with a certain level of profit, Twelves was adamant that the work itself, and the workers' satisfaction, ought to guide the enterprise.[26]

Whether she was entirely correct or not, the result was that Twelves decided to leave the Langdale Linen Industry and move to Keswick to join the Rawnsleys, who had set up the flourishing Keswick School of Industrial Arts (KSIA). There she headed up the linen department. In just five years, she established extremely high standards and won the patronage of "Princess Louise, the duchess of Albany, Lady Muncaster, Ruskin and Joan Severn." Yet, by 1894, Twelves was at odds once more with her social superiors. The KSIA had appointed a central committee on which Mrs. Rawnsley, herself a savvy and talented woman, played a significant role, along with other volunteers who came from the wealthier class. None would have possessed the skill and direct knowledge of Twelves herself, but this committee oversaw Twelves and her students. Grievances "seem[ed] to be centred around Twelves' concerns about individual effort and control." Once again, Twelves felt compelled to leave. Some connection was maintained between Twelves and the KSIA, though ultimately she broke away because she felt her expertise and insights were not thoroughly acknowledged by the KSIA's central committee. It seems that the patrician leadership in the Arts and Crafts movement had difficulty ceding power to their social inferiors, no matter what their talents.[27]

Nevertheless, Twelves went on with her linen work for thirty more years as head of the newly established "Ruskin Linen Industry" under the jurisdiction of Ruskin's Guild of St. George. The KSIA also did well in wood carving and metalwork, but its linen industry effectively ended after Twelves left. By contrast, the Langdale Linen Industry continued for many years under the guidance of Elizabeth Pepper, who kept alive the ideals and skills cultivated by Fleming and Twelves. Perhaps her most extraordinary contribution was a project to rear silkworms in local mulberry trees. This strange scheme was a logical if utterly impractical variation on Fleming's original plan to make truly local textiles.[28]

THE PUREST BOOK

The aim of Fleming's linen manufacture was to channel Ruskin's thinking into an accessible public experience and to make people see the world anew. He wanted his textiles to remind the consumer of the chain of labor and raw materials that made each object. Each handmade piece of clothing showed the buyer how consumption could keep tradition alive and preserve cherished landscapes for posterity. This kind of social vision has become a little more common in our own time, but for Fleming and his contemporaries it was a radically new way of seeing the world of goods.

Nowhere was Fleming's idealism more evident than in the little book made to celebrate the Langdale Linen Industry by his ally H. H. Warner in 1889. Warner compiled dozens of famous poems from ancient times to Wordsworth's day on the beauty and wonder of the craft of hand spinning and weaving. He called the collection *Songs of the Spindle & Legends of the Loom*. The book extolled the pleasures of handicraft production and consumption. Warner pointedly included, "as far as possible," the names of "all *craftsmen and workers* concerned in producing this volume." Labor ought to be a concern for everyone. "[T]he toiler" was "best helped 'by a right understanding on the part of all classes of what kinds of labour are good for men.'" The proper kind of skilled work, he emphasized, ought to eschew the "cheapness" of machine-made goods while at the same time creating "'demand for the products of healthy and ennobling labour.'" Workmen should also make products from materials that would keep nature at the forefront of the consumer's mind. On a visit to the British Museum, Albert Fleming had come across an ancient piece of cloth, "more than thirty centuries old." He found it wonderfully preserved, "unequalled for beauty of texture." Modern-day industrial textiles were completely ephemeral by contrast. It was impossible for him to think they would survive long enough to be collected in museums. Even if such an item *were* to survive, viewers would have a very different experience of the piece as they imagined the conditions under which it was made. Machine-woven cloth from the period could hardly reflect intergenerational bonds between mother, daughter, and grandchildren, as in the case of Eleanor Heskett (who may have taught Marian Twelves to spin), her daughter Elizabeth Pepper, and Pepper's daughter and granddaughter (see figure 2.3). Factory-made cloth also lacked a sense of place. It could not call to mind the Cumbrian hills tufted with grass and scented with heather, or the long afternoons full of friendly chatter among spinners.[29]

Songs of the Spindle & Legends of the Loom was a symbol of all that had been lost, and more importantly, all that could be regained. It was handmade in almost every way. H. H. Warner's preface detailed each significant feature. In an age when consumers were agog with wonder at steam-powered marvels, Warner and Fleming hoped to force readers to pause and consider the local and individual features of the object closely. Warner observed: "The buyer of a thing may seldom think of the workers' sacrifice in producing it, yet the sacrifice of unremitting and often ill-rewarded toil should be thankfully acknowledged." His aims in producing it were to "preserve . . . individuality and human interest, as the price at which it is offered will permit." He went on: the paper was "made by hand"; it was printed on "a hand-press"; the flax from which the linen cover and the paper were made "was first spun by cottagers at their wheels in the Langdale Valley"; and the thread they made was later woven "on the hand loom at the same place." He further described the natural methods of processing the materials. The cover's linen was "unbleached" and "therefore the natural colour of the dried flax"; no "deleterious chemicals" were used, but only the "pure mountain air and sunshine." Warner was proud of its small print run; he felt that turning out "thousands of a thing, each of which is alike in detail and finish, at once diminishes its art value." Machine-made goods were "monotonous in their uniformity."[30]

Copies of *Songs of the Spindle & Legends of the Loom* are still available for purchase every now and then. The Huntington Library in Los Angeles owns a volume. Its linen cover is a dark tan and roughly handwoven. For anyone aware of its history, the book remains an object of peculiar interest. There is something otherworldly and fragile about the slim text, despite the coarse cover and warped paper. One contemporary reviewer was struck by the "pathetic" qualities of the little book. The very fact that anyone should have to "boast" about handmade qualities "as a curiosity" was "the saddest thing about it." To the anonymous observer, the book was not a symbol of triumphant revival, but plainly of loss. If such books were to be made more often, "they would, of course, have to be better done than this." Even more troubling than its aesthetic failings is the fact that this book, which was supposedly "the product of *hand-work* alone," made use of a modern innovation. While some of the images are made from woodblock prints, several were created by means of auto-gravure, a process that grew out of late nineteenth-century technology associated with photography and "rotary printing from cylinders." Further, in the list of workers included at the front of the book, the authors have left blank the makers of the paper. Warner noted explicitly in the preface that the paper was handmade, and that the flax was spun in Langdale for use in both the linen cover and the paper. Yet many of the pages bear the watermarks of Dutch companies, indicating that the paper was actually produced in Holland. The flax itself had been imported from Ireland after attempts to grow it in Lakeland failed, though it had been grown there traditionally. *Songs of the Spindle* was not able to achieve absolute purity in production then, but it was still an inspiring creation. A superior book cover was designed and produced for Joan Severn around 1898 by the Ruskin Linen Industry under Marian Twelves's supervision. Like the earlier book cover, it was handmade, but it featured elaborate, delicate silk embroidery. With its fine floral motif, it is both elegant and subtly visionary, reminding one of the power of Fleming's idealism. His contribution to the technicalities of making and selling handmade linen products is also evident in the success of Elizabeth Pepper's exhibition displays. She had, for instance, won "international acclaim at the 1893 World's Columbian Exposition in Chicago." In 1897 the Langdale Linen Industry attended the Lancaster Arts and Crafts Exhibition and "won fourteen awards." A photograph from 1906 shows Pepper at the Home Arts & Industries Exhibition in London, her table overflowing with decorated linens. One can see both the tenacious entrepreneurial spirit and the adherence to fine craftsmanship inspired by Fleming. Prominently draped on the

FIGURE 2.3 Mrs. Elizabeth Pepper with her mother, Mrs. Heskett, and her elder daughter, Mrs. Nelson, holding her daughter, Abigail Reed. Elizabeth Pepper took over the Langdale Linen Industry when Marian Twelves joined the KSIA. Mrs. Heskett's mother handled the accounts. The latter may have been "the old lady from Langdale" who taught Marian Twelves to spin, and who also spun the thread for the cover of *Songs of the Spindle & Legends of the Loom*. Thanks to Vicky Slowe for this information. By kind permission of the Ruskin Museum, Yewdale Road, Coniston, Cumbria, LA21 8DU, UK.

front of the table is an arresting example of Ruskin Lace, with its trademark square pattern. Evidently, one piece of linen bore the mantra "make new friends"—a nod perhaps to the home industry's resolve to connect social consciousness with business savvy.[31]

The latter stages of the Langdale Linen Industry and other arts and crafts associations were dominated by figures such as Elizabeth Pepper and her mother, Mrs. Heskett—by the spinners and embroiderers rather than Fleming. This was perhaps partly on account of his tragic falling out with Ruskin in 1887. It was the same year that Ruskin went to Folkestone and embarked on the exorbitant shopping spree that prompted his valet to write to Joan at

Brantwood pleading for help. Even before this event, however, Ruskin had been behaving erratically in other ways. He drastically increased all his servants' wages at a time when Joan was struggling to balance the checkbook, and he began to argue with friends, family, and even his publisher. He was not obviously out of his mind, not feverish or frightened, only volatile and unreasonable. This made it difficult to handle him, which in turn put his estate—and his future well-being—in jeopardy. After a bitter quarrel with Joan, he left Brantwood and moved into the Waterhead Hotel on the other side of the lake. In Ruskin's mind, he was being manipulated by a greedy and ungrateful family. To his credit, Albert Fleming rushed to his mentor's side, offering comfort and probably legal advice. Ruskin wrote to him days later, "I knew you would do all you could for me, but there is nothing whatever to be done—except to get the Severns out of my house." With Fleming's help, Ruskin soon got his wish. Joan was replaced by another woman—selected by Fleming—as head housekeeper. Ruskin returned to Brantwood, once again master of the house with Joan far away. Yet he felt no sense of triumph. In fact, he realized how abominably he had behaved, and he begged her to come back. After a time she did, but she had grown wary of Fleming and his legal counsel. Fleming likely imagined he would be rewarded for being a close friend, a valuable professional, and a good disciple. But just four years after the start of the Langdale Linen Industry, he found himself effectively barred from the company of the man who had so significantly shaped the course of his life. Letters from his close confidante, Susanna Beever, who as a dear friend of Ruskin also witnessed the entire episode, reveal that she was disturbed by Ruskin's refusal to meet with Fleming after this unfortunate affair. Ruskin bemoaned Fleming's unquestioning helpfulness, lamenting, "Oh, why didn't Albert see that I was mad?"[32]

Despite these conflicts and setbacks, Fleming remained actively committed to Ruskin's principles for at least another two years, submitting the "Forewords" for *Songs of the Spindle* and writing an article on the history of hand spinning and weaving. The Langdale Linen Industry itself operated until 1925. Its longevity affirms the arts and crafts values on which the whole enterprise was based—it was after all labor both loved and directed by the workers themselves. When Ruskin died on January 20, 1900, he was buried in Coniston. Marian Twelves set about making the funeral pall designed by KSIA director Edith Rawnsley and Harold Stabler (1872–1945). Ruskin had always decried the use of black for funerals, so it "consisted of plain unbleached, hand spun, hand woven linen . . . lined with rose-red silk and embroidered with wild roses." It seems that the names of all the people who worked on it were recorded, including Mrs. Youdale, who spun the thread, Robert Shearman, the weaver, and all seven women working under Twelves. In the center appeared the words "Unto This Last," after Ruskin's original 1860 work on the ethics of consumption.[33]

CHAPTER THREE

Queen Susan

There were many visitors to Brantwood in the late nineteenth century, including clergymen, scholars, university students, and artists. There were even the occasional tourists from America or the Continent. Many came by way of Hawkshead, steering their horse-drawn carriage down a narrow serpentine lane and continuing along until they came to a fork in the road at the head of Coniston Water. To the left, about a mile away, was Brantwood, with its steep and craggy hillside and dense woodlands. On occasion visitors hesitated, perhaps having heard the Master was ill or that he was busy with important guests (such as art critic Ernest Chesneau, Harvard art historian Charles Eliot Norton, or Charles Darwin). A great many surely felt bashful, not knowing the Professor personally. There was another way to enter into Ruskin's circle. They could turn right instead. Taking this route, they soon passed the old Waterhead Inn and made another quick right onto a gravel drive. As they wound their way up the lane, they came upon a large yet modestly proportioned house and its well-tended garden—the Thwaite. This was one of the favorite haunts of Ruskin in Coniston, who called it an "apple-perfumed Paradise." Another admirer observed that the Thwaite was "in conception and arrangement unlike any garden outside the realm of dreams." Such praise was not entirely unwarranted. Near the roadside grew damask plum trees and gooseberry bushes, and just beyond these in a wide open sunny space lay a kitchen garden. In four separate square beds, enough herbs and vegetables were cultivated to supply the household and share with friends. Roses, nasturtiums, and convolvulus (Morning Glory) softened the rows of edibles. Stone-edged paths led to a

FIGURE 3.1 The Thwaite, by Isaac Cooke, ca. 1891. From William Tuckwell, *Tongues in Trees and Sermons in Stones.*

"picturesque tool-house," and an "archway covered with Clematis" framed the entrance to a steep and terraced hillside garden. Throughout, the choice of flowers was pointedly "old-fashioned"—nothing exotic. Yet this was all part of its charm. After passing the pretty and practical lower garden, visitors ascended through the sloping orchard above with its apple and pear trees. This in turn was bordered by more roses and a hazel alley laced with periwinkles, "ivy-clad tree-stumps," and an "outcrop of native stone planted . . . with rock-plants." This was the property of Susanna Beever, an elderly spinster who had lived in the house since 1827. She was known both for her striking, easily accessible garden, and for her surprisingly deep connection to the Professor across the lake.[1]

In the 1870s and 1880s, Beever enjoyed quite a privileged position among Ruskin's Lakeland friends and followers. Ruskin's correspondence with her was even published in Albert Fleming's *Hortus Inclusus* in 1887. Something about the elderly spinster's life and garden had drawn the famous Oxford professor to her. In fact, Ruskin leaned heavily on her advice and friendship as he struggled with successive bouts of illness, depression, and fears of "the plague wind." Yet Beever has remained virtually unexplored in the Ruskin scholarship. The correspondence is generally cast aside as mere chatter, significant only for shedding light on Ruskin's gentler side, as he signed many of the letters as "Puss" or "Professor Grimalkin." But a careful look reveals the wide range of roles played by Beever. She was sometimes a close friend, sometimes a "mother," a "sister," or a playful "young girl of thirteen." For Ruskin, Beever represented "stability" and "an earlier age," exhibiting "moral concern" and an ability to describe nature "with delicacy and wit."[2]

Beever was a crucial figure in Ruskin's final decades, and not nearly as typical an elderly Victorian spinster as one might imagine. Ruskin wrote at least 900 letters to her, allowed her to edit selections from *Modern Painters*, and once addressed her as "Queen Susan." In *Sesame and Lilies* (1865), Ruskin's educational treatise for men and women, he developed a notion of women's special authority, what he called their "queenly office." Many critics after 1900 understood this as merely the consignment of women to the feminine or domestic realm. But as one scholar has shown, readers in Ruskin's own time, including spinsters, suffragettes, and men with unconventional attitudes toward gender, took the notion of queenliness in a more radical sense. Such readers had good reason to look for interpretations that might disrupt traditional views, and Ruskin's rich and subtle language rewarded their careful analysis. A Ruskinian Queen was always connected to the world of work, morality, and national welfare. Ruskin did not dismiss Beever as a socially "redundant" old maid, but in fact saw a great deal of potential in her. He even conferred upon her the most astonishing praise: "you know," he wrote, "you really represent the entire Ruskin school of the Lake Country." Simply put, he appreciated Beever's capacity for satisfaction in simple things and her reverence for nature. Although her outlook in old age was darkened by his dire prognostications about pollution and climate change, she became a concrete model for the ideal of the sufficient life.[3]

THE THWAITE

Susanna Beever's mother died when she was very young, and after her father suffered a serious financial setback in the early 1820s, the family moved to Coniston from Manchester. Beever (the youngest at twenty-one), her father, one of her two brothers, and three other sisters took up residence at the Thwaite. They may sometimes have felt socially isolated. The villagers of Coniston were farmers, shepherds, idle spinners (their skills superseded by urban factories), boat builders, coppice workers, bobbin makers, and slate miners. A few tourists came in the summer to enjoy the surrounding lakes and hills, but in midwinter the village could be bleak and lonely. Severe weather sometimes halted mail delivery, cutting them off from urban society still more.[4]

In 1824, just a few years before Beever arrived in Coniston, a very young John Ruskin traveled to the Lakes from London with his doting parents. Later he recalled how his experience opened his eyes to the sublimity of nature—even in its smallest parts. His parents brought him "to the brow of Friar's Crag" overlooking the labyrinth of islands covering the northern lake of Derwentwater. He remembered being overtaken by an "intense joy mingled with awe" as he looked "through the hollows in the mossy roots, over the crag into the dark lake." Ruskin's parents spared no expense on his education, traveling frequently and generally grooming their child for greatness. For Ruskin, the Lakes region was one of many stops on his life's journey. For Beever, this would be the whole of her life.[5]

Beever's father died in 1831, four years after the move. Her brother in Manchester passed away in 1840. The remaining siblings lived together at the Thwaite, all unmarried. Her other brother, John, pursued a variety of rural and artistic interests. Mary and Susanna contributed botanical specimens to several botanists and corresponded with them.[6] Mary eventually had a plant named after her: *Lastraea felix-mas* var. *Beevorii*, a type of fern.[7] Both were cited in botanical works. Gradually, the siblings acquired a certain status and esteem in the community. The family was "noted for originality," and it was said that "all were interestingly peculiar, and each in a different way."[8] By the 1830s, while Ruskin received accolades at Oxford, winning the Newdigate Prize for Poetry in 1839, Beever was developing hopes for creative success herself. She lived across the lake from Tent Lodge, home of the eccentric, astoundingly talented Elizabeth Smith (1776-1806).[9] Brantwood was for a time inhabited by William Linton, a radical, anti-industrial, republican utopian, along with his wife, Eliza Lynn Linton, a well-known journalist and novelist. Wordsworth lived just two valleys over in Rydal. Beever had tea with him and Thomas De Quincey once.[10] The Romantic poet cast his light on the rugged natural beauty of the area, prizing the vernacular over the artificial and effete. What earlier tourists had mentioned in passing with a shudder—the rocky hills and boggy moorland—had taken on a new mystical and aesthetic significance. Beever made this Romantic vision her own, celebrating the life and beauty of the rough fell lands. She wrote poetry and painted and sketched with notable skill. Like many other middle-class women in the period, she also turned to philanthropic pursuits. This was one of the few paths by which women could enter civil society without jeopardizing their reputation. She wrote two pamphlets on poor children's charitable education in 1852 and 1853 and, together with her sister Mary, contributed to a small working-man's journal.[11] In all these different ways, Beever tried to imbue her humble existence in a remote rural village with meaning and purpose.

Beever's ambitions were slow to bear fruit. In 1860 she was fifty-four and had recently lost two siblings—Anne and John. Her literary aspirations had failed to bring her public attention. The poems went unpublished. Tennyson, the new poet laureate, spent time in Coniston, but there is no record of Beever meeting him. In 1870, at the age of sixty-five, she published a translation of *King Lear* into basic prose—an attempt to render its deeper meaning accessible to the general reader. In the same year, Beever published an additional book of quotations from Shakespeare. This book lacked purpose and coherence, but she had found a literary niche in repackaging lofty works into casual formats. However, Beever's health was fragile. Evidently her mother's early death had a significant impact on her. An anonymous obituary writer informs us that "life was harsh to her in childhood, for in those days children were not pampered at school, and little girls of the wealthier classes often went hungry and cold, when they were delicate and shy, and when there was nobody to care for them." As a result, she suffered "permanent ill health" due to such hardships in her youth. A sense of slow, inevitable decline weighed on her. Her family of eight had become a family of three, and now another sister was ill. By 1873 the ending of this story seemed easy to guess.[12]

Around this time, Ruskin's future also looked dim. He had climbed to the pinnacle of fame, but he was no longer content to play the darling of the establishment. A decade earlier he had courted controversy with *Unto This Last* (published in book form in 1862), speaking fiercely against capitalism and the social decay caused by industrialization and mass consumption. The book gained a passionate following in years to come, including artists and writers such as Mahatma Gandhi and Leo Tolstoy. But as powerful as his arguments were, *Unto This Last* was a critical and commercial flop when it came out. Even ten years after publication, less than a thousand copies had been sold; a second edition was postponed. Further, he was now caught up in a romantic infatuation with the very young girl Rose La Touche. After learning that Rose was seriously ill, Ruskin suffered his first attack of mental illness, the start of a long, nightmarish series of recurring breakdowns. The day before the attack, Ruskin had begun work on the August issue of *Fors Clavigera*, where he announced for the first time the coming of the Storm Cloud. In the midst of this tumult, Ruskin's universe collided with that of Susanna Beever. He was offered the chance to purchase Brantwood from William Linton, the republican printer. From now on, he would spend a good deal of time in Coniston, living just across the lake from the elderly spinster.[13]

FROM THE WOOD TO THE GARDEN[14]

When Ruskin set forth from Brantwood to call on the Beever sisters for the first time in 1873, he probably rowed himself across the lake, as he often did thereafter. After tying up the boat, he must have crossed the main road, wandered past the Thwaite's plum trees, through the vegetable plots, up the slate steps, and along the zigzagging path of terraced plant beds. There were rock roses, periwinkles, daffodils, and scarlet anemones, to name a few—all of them quite ordinary inhabitants of the Victorian garden. Beever's taste in gardening was underpinned by moral considerations similar to those weighing on Ruskin. The effortless beauty of common plants appealed to both of them. Ruskin despised exotics that needed coal-heated greenhouses.[15] Another concern was the fit with local soils and climate. Ruskin corresponded with William Robinson, author of *The Wild Garden* (1870), and appreciated the new fashion for wilderness gardening.[16] Robinson argued for the use of hardy plants (even exotics) well adapted to their location. Beever's garden possessed plants suitable to the Lake country.[17] Some were native to Northern England and were locally frequent in Upper Teesdale and the Lake District, like *Gentiana verna* and *Potentilla fruticosa*.[18] Her correspondence mostly mentions plants that had been naturalized in Britain, such as scarlet rhododendrons, Herb Robert geraniums, Travellers Joy (clematis), *Lithospermums* (gromwells), saxifrage, rock roses, and sweet brier.[19] The few obvious transplants in the list—*Schizostylus* (Kaffir Lily), *Senecio pulcher* (ragwort)—were hardy and preferred the rocky soils found in the Lake District.[20] In a letter she once mentioned heliotrope (one of the least suitable plants for the northwest), but perhaps only because it was a dear old friend's favorite.[21]

Whereas her sister Mary was said to be "strong in practical insight and efficient help," Susanna was always considered the more "poetical" and imaginative of the two. When Ruskin first met her, she likened Ruskin

FIGURE 3.2 Beever's garden at the Thwaite, ca. 1891, showing the slate seat that Ruskin playfully described as "two deeply interesting thrones of the ancient Abbots of Furness." From Tuckwell, *Tongues in Trees and Sermons in Stones*.

and herself to characters from *Rob Roy*. She was the rough but cunning gardener, and he was the Master who needed someone to look after him. It was a fitting allusion, given Ruskin's love for Sir Walter Scott, not to mention his wounded psyche. She was a disciple, but one with a mind of her own. An acquaintance observed that the two "became intimate at once." Paradoxically, this friendship deepened when Ruskin traveled and they had to resort to correspondence to stay in touch. Beever confided in a letter that she felt she had known him all her life. She wrote, "at Coniston there is no one who *really* suits me, save you & Joanie." She complained of not being understood by her other friends, "For I am such a very strange being & so unlike people in general, that I do not even understand myself." It was Ruskin rather than her family who grasped the truth about her life.[22]

Beever's satisfaction with simple pleasures was very appealing to Ruskin. She spoke of her need to be in the garden every single day. A vase of cut flowers looms large in the only portrait we have of her, painted in 1892 by W. G. Collingwood. Wearing a large white flounced bonnet and a knitted shawl with a brooch, she is surrounded by curious objects, including a geological specimen, an owl figurine, and a painted owl—all possibly gifts from Ruskin or his followers. In later life Beever was known as the "Owl of the Thwaite." Her love of birds and plants, expressed in lighthearted sketches and vignettes, acquired a deeper meaning in Ruskin's brooding mind. He had long been dismayed by the "cruelty and ghastliness" in the theory of evolution and longed for a natural world unsullied by such notions. Beever's domesticated scenes were devoid of harsh realities and death. Her sparrows squabbled in the ivy "like many human beings," but they were safe in this *"fashionable* place of resort," and free to tell each other "the events of the day." Beever kept a hermit crab in a tank and observed it eating by "holding a tiny mutton collop in one hand & tearing fragments off with the other." Rather than finding nature repellent, she likened the scene to a boy eating gooseberries—"very entertaining." This light way with words made it easier for Ruskin to put aside his darker thoughts from time to time and to feel a sense of wonder in nature again. In what may have been an underhanded rebuke, Ruskin once showed Charles Darwin some of Beever's sketches as an example of "the true old school of drawing."[23]

Beever was modest and quaint on the surface, but at times Ruskin encountered a more commanding presence. She knew he was proud and sometimes touchy; she was also thoroughly aware of Victorian patriarchal ideals. She had to be careful about dispensing advice. Sometimes she overstepped. Once she suggested a number of "old fashioned" plants for Ruskin's new garden, unexpectedly unleashing his ire. She apologized profusely: "when you said, (rather awfully—) 'If *you* will manage *my* garden' I was in the Valley of humiliation . . . & felt too that I had *seemed* officious." When they reconciled, Beever made the cheeky reply that perhaps she *should* manage his garden: though "*not* . . . until I know what wages you give." She knew she had a great deal of experience and knowledge to offer. On occasion, Albert Fleming likened her to both St. Francis and Henry David Thoreau. In a letter from 1873, she once again gave unsolicited gardening advice to Ruskin while he was abroad, advocating an exceedingly light horticultural touch:

> Do you ever send home orders about your Brantwood? I have been wishing so much that your gardener might be told to mix quantities of old mortar and soil together, and to fill many crevices in your new walls with it; then the breezes will bring fern seeds and plant them, or rather sow them in such fashion as no human being can do. When time and the showers brought by the west wind have mellowed it a little, the tiny beginnings of mosses will be there. The sooner this can be done the better. Do not think Susie presumptuous.

Despite the mild and deferential tone of Beever's many letters, passages like this reveal her confidence. Although Ruskin had much to teach her about plants discovered in his wide travels, he also happily sought out her advice (and Mary's) as he embarked on different experiments at Brantwood. He asked in 1876, "Please, can your sister or you plant a grain or grains of corn for me, and watch them into various stages of germination? I want to study the mode of root and blade development, and I am sure you two will know best how to show it me." At another point, he asked Beever to tell him what the seed he sent her would hatch: "I'm frightened to plant it." She or Mary sent him at least one specimen for *Proserpina*, and he praised all that he learned from them, not to mention Susanna's excellent Latin. The sisters also seemed perfectly in tune with their "invaluable" gardener, Harry Atkinson. By contrast, Ruskin complained about his own gardeners, who contravened his ban on hothouse plants: "I've to rout the gardeners out of the greenhouse, or I should never have a strawberry or a pink, but only nasty gloxinias and glaring fuchsias."[24]

THE SUFFICIENT MUSE

Hardwicke Rawnsley once referred to Beever as Ruskin's "lover"—likely in a teasing, lighthearted sense. Yet Beever's role was more important than that of soothing correspondent and close friend. If Ruskin's theory of life as wealth was to have any teeth, someone would need to prove in practical terms that a simple life in the country was not just a series of deprivations suffered by the poor or those otherwise constrained; nor should it only be enjoyed temporarily by wealthy tourists. Someone had to live there permanently and consciously appreciate that life every day, for decades, without serious regret. They needed to be *satisfied* with this life. It was in fact a tall order. For his own part, Ruskin frequently went away; he found reasons to take long sojourns in London, Oxford, and continental cities more often than not. Besides, he once confessed that living too long in the country made him less able to appreciate nature; time away was required to "restore the old childish feeling." Even Wordsworth was said to have "lost the power to be impressed" if he spent too much time anywhere. In order to apprehend the merits of this way of life, Ruskin needed to leave and return periodically to see it with fresh eyes. This gave him ample reason to value Beever's contentment and wonder. If no one of a certain status, education, and means actually *embraced* the sufficient life, then it became rather difficult to defend the idea of an alternative to consumer society.[25]

Ruskin's ideal middle-class women were expected to play various roles, although these roles were not as rigidly defined as many have guessed. In Ruskin's own time, readers understood his proscriptions for the education and activities of women to be complex and variable. One model of behavior was based on St. Ursula, who could be described as "a perfect . . . helpmate: virtuous, learned, chaste, capable, and industrious, but submissive and self-sacrficing . . . a leader, a princess." Female members of the Guild of St. George were encouraged to "set an example in frugal living, avoiding luxuries and wearing plain clothes." Unmarried women of a certain class could pursue work that required knowledge and skill, serving as "'delicate gardeners . . . nurses—readers—Teachers . . . housekeepers, the most useful members of all the society.'" Beever, who was a member of the guild, fit this latter category to a great degree, although she never worked for a living. In truth, Ruskin's "ideal woman" was a constantly evolving concept. Her location in Coniston and her proximity to Ruskin gave her a privileged standing in the Lakeland Arts and Crafts circle as well.[26]

On his visits to Coniston, Ruskin could not have failed to notice the Thwaite's mix of rural isolation and cultured taste—a perfect laboratory for practicing his ideal. Beever was genteel enough that she had been told never to stoop—not even in the garden; instead, she attempted to weed by using her foot. Yet her lifestyle was rather plain; the food she mentioned in her letters was hearty, uncomplicated fare that was often grown on site: marrows, peas, cabbages, melons, apples, pears, damsons, and elderberries. The Beever family likely kept some poultry, including bantam cocks and grouse. Ruskin begged her once for some rosemary and lavender, and thanked her for sending oranges and "brown bread" and cranberry plants to set out on his moor. Such a quantity of vegetables arrived at Brantwood one year that Ruskin complained: "Only please now don't send me more asparagus!" Some edibles were in fact too delicate for the northern climate. Beever noted that the oranges she could get in Coniston were not very good, and her melons must have been started under glass. The cranberry plants she gave Ruskin no longer grow where he planted them.[27]

In any case, consuming local food was only one aspect of a satisfying, simple life. A variety of skills and forms of knowledge were needed to serve the common good. Beever had learned a good deal about the value of practical experience from her brother John. His book, *Practical Fly-Fishing*, explained the superiority of handmade fishing rods and the virtues of local observation—for instance, understanding what types of flies best attract specific fish. He valued knowledge, no matter how uncouth or poor the source, and he was proud to have learned excellent new techniques from clever working men. The Beever family also kept a fish pond behind the house for many years, developed a hand press, and printed Susanna's poems and other materials for the local school. They were aware of how fortunate they were. While the Beevers fished largely as a diversion, the rural poor around them did not have such a choice. Vagrants roamed the countryside, sometimes stealing from gardens or fishing for subsistence. Beever disapproved strongly when commons and lakes were closed off. Suddenly poor men were required to purchase a license to fish. It was a "paltry & cruel" form of oppression since fishing "used to be such a pleasure to a weary & overworked man." She believed that performing physical labor—producing a handcrafted fishing rod, or even fishing itself—was essential to both body and soul, for rich and poor alike, and she worked to make the lives of those less fortunate easier. Her obituary notes that "her special department was 'doctoring.'" The Thwaite was known as a "free dispensary," and she "became practically a local assistant" to the doctor at Hawkshead; she even provided "surgical help" to him, proving herself game and levelheaded as well.[28]

The sufficient life was not confined to basic needs. Educated people had a duty to appreciate the finer things in life as well. This required an understanding of how these things were produced. Beever had been introduced to artisanal crafts early on, at least since her brother erected a lathe propelled by a water wheel: "He used to turn all sorts of pretty and curious articles, to carve—long before the days when wood-carving came into fashion—and to make elaborate inlaid mosaic of ingenious design." She later supported the hand spinning and weaving of the Langdale Linen Industry and took up spinning herself. Beever at times showed an uncommon attentiveness to Ruskin's conviction that wonderful objects depended on the skill and contentment of the maker. When he sent her a gift of a carefully wrought inkstand one winter, she thanked him by cherishing every detail of the object at some length: "What a joy there must be in producing beautiful things—this reminds me that God looked upon His people and behold they were good!" Her gratitude was not merely a sign of etiquette, but of moral judgment. The works of craftsmen, unlike machine-made goods, were almost sacred to them both. Beever's letters indicate that she suffered from arthritis and so probably no longer attempted challenging forms of artwork herself. But she appreciated Ruskin's efforts at drawing and painting. Their correspondence reveals that she often sent him the feathers of birds, particularly peacocks, so he might study the plumes more carefully. In turn Ruskin wrote about

FIGURE 3.3 A peacock feather drawn for Susanna Beever bearing the inscription "For Miss Susie, To show how her spoiled pets dress. J. Ruskin, Dec. 7th, 1873." Reproduced by kind permission of the Collection of the Guild of St. George, Museums Sheffield.

them in *Love's Meinie* and *The Laws of Fésole*. He made drawings of them and presented them as gifts to Beever or other friends. Her reactions must have pleased him, since he heaped lavish praise on her: "What infinite power and treasure you have in being able thus to enjoy the least things, yet having at the same time all the fastidiousness of taste and fire of imagination which lay hold of *what is greatest in the least,* and best in all things!"[29]

In remote areas, simple living depended on communal economies; exchange was crucial. One did not expect to buy everything from shops. So, for instance, plants and seeds were exchanged among Beever, Ruskin, Reverend William Tuckwell, his wife Rosa Tuckwell, and others. Food also circulated in this gift economy; sometimes it was the "brown bread" that Ruskin was fond of, or fruit and vegetables. At other times it was a gift of homemade "cream cheese in green leaves." Whatever their purpose, they were valued for pleasure and variety as much as plain utility. Indeed, exchange extended also to intangible goods like useful information or stories. Susanna's brother John was a great storyteller who delighted his family with his yarns. Susanna told stories too, but hers were related to friends outside the family. Her letters are full of tales about cats, dogs, and birds and their curious habits. Ruskin and Dr. John Brown both admired her descriptive talents. Storytelling connected listeners to the natural world and the history of the land and its people, encouraging meditation on their well-being too.[30]

FRONDES AGRESTES

In 1874 Beever's sister Margaret was dying. Soon she would have only Mary left. Beever wrote to Ruskin in great sadness, "I am astonished to find myself 68 . . . Much illness, & much sorrow, & then I wake up to find myself old—& as if I had lost a great part of my life." Yet she still hoped that it was not "all lost." Ruskin's response was heartwarming. He knew she had been collecting her favorite passages from *Modern Painters*, much as she had done with Shakespeare. He encouraged her to gather and publish them, intending it partly as a helpful distraction. Beever must have wondered at the thought that her name would be attached to Ruskin's. What would people make of her selections? Would any of her own feelings and thoughts show through? Anthologies compiled by men during this period were often accompanied by explanatory and evaluative remarks, whereas those compiled by women usually contained little or no commentary. The selections of women were considered mere "pearls of wisdom." But one scholar has argued that women editors of Ruskin anthologies conveyed their analysis in a subtle manner, since from the trove of Ruskin's writings they chose specific quotations that inspired them or which they hoped would influence their readers. In fact, Beever's selections did just that. In a sense they highlighted her own experience of the simple life.[31]

Ruskin advised Beever to "think of the form the collection should take." He described as a "reference" point his own plans for a new edition of *Modern Painters* in which he would "take the botany, the geology, the Turner defence, and the general art criticism . . . as four separate books, cutting out nearly all the *preaching*, and a good deal of the sentiment." Meanwhile, he asked her to focus on the *"didactic* . . . as opposed to the other picturesque and scientific volumes." Despite his somewhat conflicting advice and impulse to direct the project, Ruskin obviously assumed that Beever's book would be valuable in any case because of *who* she was: "Now what you find pleasant and helpful to you of general maxim or reflection, *must* be of some value." Ruskin emphasized that he trusted Beever's judgment, telling her, "I mean to leave it *wholly* in your hands." She was "exactly in sympathy with [him] in all things." Nevertheless, he was frustrated by her suggested title, *Word-Painting*, since he now disliked the long-winded, descriptive prose he had formerly cultivated: "I thought it was the thoughts you were looking for?" He chose the title *Frondes Agrestes* himself. As he explained to Beever, "Agrestes means what Scott means by 'wild wood'—the leaves of trees that grow at their own pleasure, as opposed to cultivated fruit trees, orderly poplars or elms,—or cared-for parks." He intended for her to have complete control so that the selections could grow in their own way. Yet he complained to his publisher that her excerpts were "sugary stuff which I can't let come out again without lemon juice." His solution was to add footnotes to what he felt were youthful errors of judgment—passages Beever still found valuable. He even cut out one quotation "about everything turning out right," unable to stomach his former optimism. In the preface, he allowed himself a few condescending comments, writing that whatever "such a person felt to be useful to herself, could not but be useful also to a class of readers whom I much desired to please." Similarly, in a letter in 1881, he described *Fors Clavigera* as "not meant for girls," while Beever's selections in *Frondes*, as well as "the new *Stones of Venice*, the Bible of Amiens, etc.," were more suitable for them. Despite these words, however, he wound up giving *Fors Clavigera* to the Whitelands College for women anyway—but not without serious thought.[32]

A closer look at precisely what Beever culled from *Modern Painters* indicates that it was more than just a young lady's book. Instead, it presented a succinct picture of Ruskin's ideal of sufficiency, calling to mind some important themes from *Unto This Last*. Beever's selections expressed in particular an appreciation for nature. She gathered long, descriptive, or even "purple" passages organized around natural themes—"The Sky," "Streams and Sea," "Mountains," "Stones," and "Plants and Flowers." But, as if to underscore the link between sufficiency and natural history, she also chose many excerpts pertaining to the ethics of consumption. In section I, on art, she quoted Ruskin on the idea that one should not seek inspiration in "gilded palaces . . . [and] artificial mountains." The passage referred to in *Modern Painters* mentioned specifically the disappointing tendency of inferior artists to be dissatisfied with scenes that were ready to hand—in nature or one's everyday surroundings. Artists and observers should not "abuse" the faculty of sight with such "selfish and thoughtless vanities." By giving in to this urge, "we pamper the palate with deadly meats, until the appetite of tasteful cruelty is lost in its sickened satiety, incapable of pleasure unless, Caligula like, it concentrates the labour of a million of lives into the sensation of an hour." One should rather choose "humble and loving ways" when seeking "the highest pleasures of sight." This was not just a technical matter for artists, but also a moral challenge for all people. By quoting this passage, Beever wanted to stress the therapeutic and moral effects of nature and the need to rein in base desires. The section "On Moralities" described the gratifying powers of "rest" not as apathy, but as "a longing for renovation" and a desire to prepare for permanence and perfection rather than a constant stream of temporary states. This kind of rest was a goal not just for artists, but for the public at large. It may have been a criticism of the modern compulsion to work too much and the preference for speed over more measured forms of satisfaction. To the woman who was told she could not weed her own garden because she was not permitted to stoop, it must have been liberating as well to quote Ruskin on the notion that "serviceable" "physical exertion" could be a great benefit to the upper classes: "It would be far better, for instance, that a gentleman should mow his own fields, than ride over other people's." The section "On Education" especially touched on the need to be content with the simple life, at the same time acknowledging the difficulty of doing so: "It is the curse of every evil nature and evil creature to eat and *not* be satisfied." To achieve this sufficiency, Ruskin wrote, "it is necessary *fully* to understand the art of joy and humble life." He meant by this that people should seek pleasure in "the loveliness of the natural world." Beever's frequent expressions of delight in nature and small amusements, along with her pointed selections for *Frondes Agrestes*, suggest a keen awareness of the character of her own life.[33]

One scholar has argued persuasively that many of Ruskin's women anthologists chose passages that highlighted the role of women as significant economic agents. In fact, *Frondes Agrestes* is not unlike Louisa Tuthill's *The True and the Beautiful* (1858), which also included lengthy excerpts on Ruskin's ethics of consumption. Yet Tuthill's book is a hefty affair with many other subheadings, making it at once more comprehensive and less appealing to the average reader. Passages on consumption hold no special weight here among dozens of other issues. Rose Porter's large *Nature Studies* (1900) contains only one or two passages related to consumption. Other women anthologists included Ruskin's injunctions against spending thoughtlessly on women's attire, quoting from works that address women directly, such as *Sesame and Lilies* or *The Ethics of the Dust*. Beever's selections, culled solely from *Modern Painters*, are not specifically related to women's consumer choices, but to a broader moral framework. In contrast with other anthologies then, her carefully chosen excerpts take on a special significance.[34]

At the time of publication, Ruskin's sharp self-criticism and patriarchal cast of mind may have prevented him from seeing the true value of *Frondes Agrestes*. Yet, despite stubborn doubts about Beever's editorial powers, he "printed her selections in absolute submission to her judgment." It was fortunate that he did, too. *Frondes Agrestes* sold well. In the first year, a second edition was issued. By contrast, *Unto This Last* had been twelve years out of print, having sold fewer than 1,000 copies. By 1900, a total of 34,000 copies of *Frondes Agrestes* had been sold—one of Ruskin's bestselling books and a "figure surpassed only by that for *Sesame and Lilies* with 40,000 copies." Collingwood mentioned in 1893 that after eighteen years in print, Beever was still earning about £70 a year, perhaps £1,500 over the course of her life. Thanks to Beever, the ideal of sufficiency had made its way to a larger audience.[35]

CHOOSING THE GOOD LIFE

By editing *Frondes Agrestes*, Beever showed that she knew something of Ruskin's political economy and moral philosophy. She was able to assert her own will and intelligence by gathering together quotations that spoke to values specific to Ruskin's circle. Even so, it is fair to wonder: just how remarkable or unusual was Beever's life compared to others living in the region? She supported the Langdale Linen Industry and took up spinning herself, despite her age, but she was nowhere near as involved as Marian Twelves or Albert Fleming. Although a decisive answer may be elusive, several clues help flesh out the nature of the choices she made. Beever had opportunities to engage in newfangled forms of consumption, but did not pursue them with much enthusiasm. Mail-order catalogues selling exotic plants were doing brisk business, but Beever's letters mention only two orders she made from a nursery in Hyères. Far more often she commented on the common plants that already surrounded her. The Thwaite gardens were humbler and more local than they might have been. Exoticism was aligned with financial wealth and social status even in the Lake District. The massive Conishead Priory in nearby Ulverston boasted extensive landscaping with exotic specimens and an "American garden," but Beever seems to have been unimpressed. Her family was educated and had the means to expend time, energy, and money on the diversions of the city, yet no mention is ever made of any visits to department stores or spectacles at the theater or the circus. She would have been in good company—Ruskin himself was glad to visit the Crystal Palace and the zoo. There is no strong evidence that she ever asked Fleming, Tuckwell, or Ruskin to bring back any special items from London or elsewhere. Ruskin expressed admiration for Beever and her sisters, who "did not travel . . . did not go up to London in its season . . . [and] did not receive idle visitors."[36]

Beever's choice to live simply is even more significant, given that she did not always find life in the country easy. Coniston suffered from the market forces of the city. She complained to Ruskin that the price of butter was high and that it was now being shipped off on the railways to places where it made a greater profit. Her outspokenness against animal abuse suggests that Coniston's farming and mining community had little compassion for helpless creatures or the people who cared for them. Beever was so empathetic that she thought it "devilish" when eggs were replaced with stones in the nests of pigeons and chickens. Few people in the village would have understood such sentiments. Rural isolation affected her social life, too. A spell of foul weather made her feel as bad as "when there is no one in Brantwood & I have no one to tell my thoughts to." She relied on correspondence to maintain close relations with friends. If the trickle of letters was interrupted, she came close to despair: "Oh my so valued Friends are you never going to write to me again?" she complained to William and Rosa Tuckwell. "I have waited and waited and hoped and hoped but it 'cometh not.'" Yet the longing for contact was mixed with a considerable measure of social reserve. In one letter to Fleming, there was even a hint of agoraphobia. She woke up "painfully nervous" one morning and was so agitated that she "did not dare to go into the drawing room!" Beever was perhaps distressed by the absence of the right *sort* of people. She admitted to avoiding those who did come at times.[37]

Ruskin knew of these retiring tendencies in Beever. He once lamented in a private letter that her talent as an artist had gone uncultivated: "You might have done anything you chose, only you were too modest." But he did not assume her shyness—or her age—defined her. Indeed, he praised her as "Queen Susan" and a model representative of "the Ruskin school of the Lake District." The idea of the "Queen" surfaced in *Sesame and Lilies* in 1865. As noted earlier, Ruskin has sometimes been decried as misogynist and outmoded in his treatment of gender. But a closer look at his definition of a "Queen" reveals the surprising flexibility and breadth of the term. Although Ruskin may have generally appreciated the notion of "separate spheres" for men and women, he also encouraged women to venture outside the home. A true "Queen," Ruskin wrote, should be the "centre of order, the balm of distress, and the mirror of beauty" "within her gates." But she was "also to be without her gates, where order [was] more difficult, distress more imminent, loveliness more rare." Ruskin's Queens were supposed to offer assistance in the unlovely, unloved parts of the world, in the slums and the poorest cottages: "So a woman has a personal work or duty, relating to her own home, and a public work or duty, which is also the expansion of that." As one historian argues, "In Ruskin's hands, 'separate spheres' was an unstable ideology." The London radical Octavia Hill is a striking example of a woman who lived and worked in both spheres. Ruskin funded her schemes to improve housing for the poor, and she made a point of observing slum dwellings in person. She also went on to become one of the founding members of the National Trust. In fact, Ruskin encouraged and financially supported many painters, translators, and reformers—many of them "spinsters" or women who fell outside typical social roles. Why was Ruskin so willing to weaken the boundaries of the two spheres? Perhaps he had good reason to sympathize with spinsters and other outcasts. Ruskin had one failed marriage behind him—annulled due to non-consummation. His bride, Effie Gray, told her father that "he had imagined women were quite different to what he saw [she] was." Ruskin never established a sexual relationship with anyone, as far as we know. One scholar emphasizes that Ruskin was an adult of ambiguous sexual—and thus social—status. The renowned professor understood quite well, then, the burdensome constraints on spinsters' lives and was sensitive to their hidden potential.[38]

Ruskin's ideal woman in *Sesame and Lilies* was partly reflective of a type that had become well established in any case. Since the 1820s, spinsters in particular had found that philanthropic activities offered a loophole by which to escape from stifling, unfulfilled domesticity. Philanthropy became their profession. Beever fit this role too; at least once, she ventured outside the safety of her circle of friends and neighbors. The British Library holds two pamphlets written by Beever in support of Dr. Guthrie's "Ragged Schools" of Edinburgh. These were charity schools set up for orphans or children of parents who were abusive, incarcerated, or had been transported to the penal colonies. Dr. Guthrie claimed that "his Ragged School [had], during the five years of its existence, saved the country nearly 100,000 pound[s]." Beever was moved to write her pamphlets after she visited Dr. Guthrie's school in 1851 (when she was forty-five), where she saw former street urchins now clothed and nourished and taught useful skills. It was a significant endeavor at this time for a respectable single woman to tour the wards of the poor in a large city. Her public appeal for money for this experimental institution was also progressive. Thus, in the manner of one of Ruskin's "Queens," fourteen years before he wrote about them, she had once ventured "without her gates" to help those less fortunate.[39]

But when Ruskin called her "Queen Susan" the same year that he reissued *Sesame and Lilies* (1882) "at the request of an aged friend" (likely Beever), he was acknowledging more than her philanthropic activity. For a true "Queen" also had to embrace ethical consumption. Ruskin had called attention to the ways in which women's consumer choices affected workers' lives when he described the making of glass beads in *The Stones of Venice*. What one bought contributed to a culture of inhuman labor conditions. The environment was also at stake. Ruskin returned to this idea in *Sesame and Lilies*. He asked his female readers to imagine private gardens, "at the back of your houses . . . large enough for your children." Then he added a diabolical twist: "[suppose] if you chose, you could double your income, or quadruple it, by digging a coal shaft in the middle of the lawn, and turning the flower-beds into heaps of coke." Would they do it? If his readers had any doubt what he was really arguing, he added that in fact, "this is what you are doing with all England. The whole country is but a little garden." In this way, he showed how the "separate spheres" of the household and work overlapped. What went on in the private sphere affected the nation and the environment, and vice versa. Only a true "Queen" understood that everything depended on coal fields near or far.[40]

THE PLAGUE-WIND

Beever only gradually seems to have become resigned to the grim possibility of Ruskin's Storm Cloud. One of her selections for *Frondes Agrestes* insisted that Nature's beauty, particularly the sky, is "all done for us and intended for our perpetual pleasure." Ruskin was quick to snap at her choice. He must have been peeved by the mention of such sunny idealism and more annoyed still that Beever had missed the darkening strain of his thought. He inserted a footnote to explain his irritation: "At five-and-fifty, I fancy that it is just possible there may be other creatures in the universe to be pleased, or,—it may be,—displeased, by the weather." This was a veiled reference to his worries about environmental decline. He had written publicly about the "plague wind" in *Fors Clavigera*. Beever would likely have encountered the idea there. Moreover, she was personally confronted with his grave suspicions in a private letter. In May 1874 Ruskin wrote to her from Rome to say, "There is nothing now in the year but autumn and winter. I really begin to think there is some terrible change of climate coming upon the world for its [s]in, like another deluge." By 1875 Ruskin was pressing his fears of climate change and pollution still more on her, complaining of Coniston's "dry black 'London-best' fog—Is it not a new, or at least a late—curse on our modern England, Susie?" Despite the sheltered place of Coniston in the northern hills, in the summer of 1879 "diabolic clouds" sent black, fitful winds to distress the vegetation. The roses in Ruskin's garden decayed into "brown sponges," and the strawberry stalks rotted. Four months later he wrote of "an unbroken fog" over Coniston—much like the London Peculiar. The edge of the lake was "covered with black scum, deposit from Manchester." Every gardener must deal with bad weather, but Ruskin suspected the "plague wind" stemmed from altogether more malevolent forces. As one scholar notes, the blast furnace at Barrow and other nearby towns spewed out soot into the atmosphere, "dimming the sun and often changing the color of the sky above Coniston and even the color of the hills." Ruskin's madness did not cause his fears of climate change, but weather changes seem to have exacerbated his illness.[41]

Beever's letters were more often than not a relief, as when she wrote soothingly that Nature, rather than technology, should be admired for its efficiency, since "mother earth . . . makes such good use of everything." Nature was "delightfully unlike most economists,—the very soul of generous liberality." Though her letters were not free of naïveté, it was perhaps what Ruskin wanted to hear. He himself was "ashamed" to doubt that the seasons would continue to come in due course. Yet whereas others regarded his fears as an effect of illness, Ruskin thanked Beever for taking him seriously:

> This July is very like the one which frightened and fretted me into my illness at [Matlock] it's a great comfort to me to have you to grumble to now—nobody believed me then, but said it was all "natural"![42]

In truth, she found his black moods oppressive. For some time, Beever had only vaguely echoed Ruskin's thoughts, musing uneasily on the bad weather. "I'm so sorry," she told him, "this weather tries you so much—The gloom is very depressing." Around 1880 he had occasion to apologize to her, using rather suggestive language: "Darling Susie, I heaped ashes on your head yesterday, and—fresh cinders into your pottage, and was a monster and a wild cat, and a serpent and a wolf and an owl and a bat and a demon." He had burdened her with his pessimistic mood. He went on to say he *had* written comforting things lately, "though I can't comfort myself—and I'll come often to be lectured." His periodic breakdowns gave her reason to doubt the reality of the "plague wind." A mocking tone sometimes colored her letters to Fleming: "Alas, what tempestuous weather it is with him!" When she fretted about unusually cold temperatures, she quoted Shakespeare, linking present variations to ordinary changes in the past: "'The air bites shrewdly' the seasons seem to have changed—but it is nothing new—Shakespeare said the same things long ago." A sarcastic complaint to Fleming, "I am quite sick of Brantwood weather," again downplayed Ruskin's fears.[43]

Ruskin's breakdown in 1887 brought the crisis to a head. Beever corresponded with Fleming about Ruskin's erratic behavior and the uproar it caused at Brantwood. She was particularly distressed by Ruskin's expulsion of Fleming from his inner circle, as well as his unaccountably heartless treatment of a pony. She and Ruskin had always shared a love of animals. Beever's obituary noted that she spoke up "against any approach to cruelty or even neglect," and stressed that concern for animals was a key part of her friendship with Ruskin. He had resigned his post at Oxford in protest against animal vivisection and had taken on the role of president for a fledgling animal welfare association. So, when he inexplicably gave the pony to people she feared would abuse it, she asked him to reconsider. She reported to Fleming, "His answer is malignat [*sic*] cruel to a degree! Alas I feel as if I could beat him!" The spat came, awkwardly, just before the publication of *Hortus Inclusus* (the collection of friendly letters between Ruskin and Beever). Ruskin's bewildering behavior alarmed her, but she held her tongue. When he eventually recovered, she was relieved to hear he felt better, even though he ignored her. Fleming asked her to chastise Ruskin, but she refused, fearful of losing "one of the greatest pleasures of my little life!" In time Ruskin mended his friendship with Beever and Fleming, and expressed gratitude for Fleming's work on *Hortus Inclusus*.[44]

A GARDEN ENCLOSED

Many reviewers of *Hortus Inclusus* found Ruskin and Beever's letters to each other embarrassingly intimate, childish, and trivial. But one of the more favorable reviews came from Arthur Galton at the *Century Guild Hobby Horse*, "a distinctly Ruskinian journal." Galton seems to have grasped how the idea of simple pleasures and friendship lay at the heart of the collection. "[E]ven in the noisy, hurried generation," he remarked, "the finer harmonies of the world may be heard by those who will seek them and listen for them." In its first four months, 1,700 copies were sold, prompting a second edition in 1888. Like *Frondes Agrestes*, this slim volume helped promote Ruskin's Lakeland ideal and also offered a captivating image of the good life in the Lake country. Beever had long felt the world was watching Coniston—or rather, that it ought to be in the public eye. Ruskin complained when she gave him a "little lecture about being 'a city on a hill'"—an example for the world to emulate. He rejected the notion that he must live an exemplary life at Brantwood—"I don't want to be anything of the sort . . ."—yet they both must have known she was right.[45]

In 1891 Beever was eighty-five. Her last sibling, Mary, had died in 1883. She had a few loyal servants and her faithful gardener, Harry Atkinson. Ruskin had fallen into the final phase of his illness and all but ceased to visit or write to her. But Beever's fame from *Frondes Agrestes* and *Hortus Inclusus* had grown. She received letters and visits from all sorts of Ruskin admirers, including Albert Fleming, the Collingwood family, Reverend William Tuckwell and his wife, Rosa, the Bishop of Brechin, and many others. "American ladies" stopped by at one point; on another occasion strangers who asked to wander through her garden were each given "a lovely rosebud." She may even have had a visit from Cardinal Henry Edward Manning (who was also a friend and confidante of Ruskin) in 1887. Tuckwell wrote without too much exaggeration of "the countless men and women of note and influence" who came calling at the Thwaite—"pilgrims to the Garden Enclosed . . . grateful for all which it has taught them."[46]

Yet Ruskin's nightmarish visions of climate change and atmospheric pollution lingered in her mind. In her letters to Fleming, discussion of the weather occurred more frequently in the last years of her life. "An old man told Atkinson yesterday," she wrote, "that he never remembered such a season"; she ruefully added, "Nearly every flower [was] starved to death!" With Tuckwell, she faithfully performed the Ruskinian ideal he so admired from *Hortus Inclusus*, speaking of her daily delight in her garden. But she also touched on more serious matters. She mentioned she was reading *The Voice of a Naturalist*, *The Malay Archipelago* (1869), by Alfred Russell Wallace, the rival of Charles Darwin. This book offered her a very detailed view of a different kind of climate from that of England. She brooded more over the weather. "The atmosphere is so thick & heavy," she wrote to Tuckwell, "that it is quite depressing—you don't know what is coming." She seems finally to have considered the possibility that Ruskin was right all along when she wrote, "Some [Frenchman] thinks that changes in climate are taking place—Lapland & France & Norway &c have been warmer than usual." There were no more mocking references to "Brantwood weather" in her letters. Beever's sense of foreboding may have colored her plans for the posterity of the garden at the Thwaite. Two years before her death, Tuckwell announced her final wish that the garden be opened to the public, complete with an enclosure for "more characteristic or rare Lake plants." The request to include rare local plants may suggest a growing anxiety about the landscape. In any case, Tuckwell responded with bombastic assurance that the residents of Coniston would approve, "for the sake of her whom above all its inmates it has learned to love and reverence."[47] He believed "the vast public" would come to see "all that has ministered to Mr. Ruskin's happiness"; "Lake residents" had a "duty to the world" to keep and promote this "paradise." Curiously, Tuckwell failed to imagine that Brantwood, not the Thwaite, would be open to visitors.[48]

The idea of public gardens at the Thwaite was not far-fetched; Beever's name continued to hold a certain literary cachet for many years. Frederick Sessions mentioned Beever in his 1907 *Literary Celebrities of the English Lake-District*. No doubt the fact that Ruskin had been buried beside her in 1900 had something to do with this, along with *Frondes Agrestes* and *Hortus Inclusus*. As late as 1921, almost thirty years after Beever's death, C. L. Maxwell was so moved by the sight of the Thwaite that he sent two photos to *the Garden*, a weekly journal. One of the black-and-white photos suggests a thick profusion of shrubs and blooms. A brief note described it as "A Link With Ruskin." Maxwell added, "The white, pink and mauve Lupins were very striking," as were the "white pinks" and a "row of Monkshood"—all decidedly common flowers. These photos represent more than a mere curiosity. In his will, Ruskin "earnestly prayed" that the house at Brantwood would be open to visitors after his death for at least one month a year. But the Severn family was not swayed by his wish. They discouraged strangers, sold off many paintings and other items, and allowed the house to become "increasingly damp and neglected." Due to Ruskin's debilitating illness, his gardens had been allowed to run wild for some time. As early as 1888, "the Professor's Garden was abandoned and returned to nature." Joan Severn kept up her own gardens on the estate for many years, but these never really evoked Ruskin's ethos, often featuring "flamboyant exotics . . . for colour, scent and dramatic effect." As Brantwood and its gardens slowly fell into

FIGURE 3.4 Ruskin's grave and tombstone in St. Andrew's churchyard, Coniston. The stone was carved by H. T. Miles of Ulverston, from W. G. Collingwood's design. The graves of Susanna Beever and two of her sisters lie to the left; on the right are the graves of Joan and Arthur Severn and their daughter Lily Severn, and beyond them are the graves of Robin G. Collingwood, William Gershom Collingwood, and Edith Mary "Dorrie" Collingwood.

disrepair, and as the Severns ceased to allow access, Beever's garden at the Thwaite may well have been the closest one could get to Ruskin's hope of an enclosed paradise in Coniston.[49]

THE OVERGROWN PATH

Little remains of Beever's own garden today. The land now belongs to the owners of Thwaite Cottage, rather than the Thwaite. The space has been reinvented to suit the needs of a new family, complete with a well-kept soccer field beneath a mature stand of deciduous trees and evergreens. Gardening flourishes instead nearer Thwaite Cottage, where Beever's friend Harriette Rigbye lived. Still, one can just make out the remains of the slate foundation of Beever's glass house. Charred bricks on one end suggest that it was heated with coal—something Beever may have later regretted. The skeletal outline of terracing up the hill reminds us how Ruskin had once roamed through Beever's "mountain garden," where "the native rocks slope[d] to its paths in the sweet evening light." He was also very taken with her slate chairs, what he playfully called "two deeply interesting thrones of the ancient Abbots of Furness." The famous slate chair in the wood above Brantwood may well have been modeled on them, small as they are. Today Beever's "thrones" are covered by moss and ferns, resting behind a five-foot-square box hedge—the only plant remaining from Beever's time.[50]

The archival path is also quite overgrown. The bulk of the extant letters relating to Beever is at the Huntington Library in Los Angeles. In 2010 the library's finding aid reported that it held 230 letters written from Beever to Ruskin. Since only a small sample of her letters had been published in *Hortus Inclusus*, this was exciting to discover. But as it turned out, the letters were instead from Beever to Fleming, with a few dozen to the Tuckwells. The error in the finding aid had been made decades ago when the library acquired the collection, yet no one had deemed it important enough to check or correct. "Oh that a woman's voice could shake the world," Beever had once written to Tuckwell. She probably would not have been surprised to hear about mislabeled letters.[51]

However, the Huntington correspondence did yield one more enigmatic clue to Beever's biography. Inside the collection was part of a poem copied out in Beever's hand. The author was a Welsh writer, who in 1857 published a book of devotional poetry. Of all the poems in the collection, Beever chose the one whose theme is the struggle to be satisfied. It is fraught with existential doubts and a brooding sense of the natural world. The "moonbeams" in the poem are not soft or bright, but "silver lances"; the frost falls "like cold words on the warm hearted." Darkness carries no relief: "the yew trees gloom against the somber skies"—"the night is oft a messenger of death." The speaker's sense of alienation is palpable and coupled with despair at not ever being heard. She protests to an angel that she "cannot sing a truth inspiring song/if none on earth will listen." The angel makes a gentle reproach:

> If there be none to listen to thy song—
> No ears to heed—no loving eyes to glisten—
> God's little Wood Birds sing the whole day long—
> And care not who may listen.[52]

Beever has sometimes been seen as a merely lighthearted, elderly companion who comforted Ruskin whenever he worried about the crueler Darwinian aspects of nature. But the fact that she copied this particular poem out and quoted it at least twice to friends in later years shows that her sunny outlook was something that she, like the speaker of the poem, had to cultivate. Her pleasure in simple things was a sign of resilience in the face of nature's harsh realities.

CHAPTER FOUR

Taming the Steam Dragon

On Christmas Eve 1874 a wheel failed on the first car of the Great Western Railway train as it passed Shipley, just north of Oxford. One passenger tried to avert disaster by waving at the locomotive from a compartment window. Yet when the driver applied his brakes without waiting for the crew at the rear end of the train to brake as well, the first car hurtled headlong into the locomotive and collapsed. Nine of the other coaches slid off the bridge at the Oxford Canal. Thirty-four people died, and sixty-five were gravely injured.[1]

A frequent traveler on the Great Western, John Ruskin found it all too easy to imagine the horror of the disaster. But the crash also reinforced a deeper aversion of his toward technology. Trains were harbingers of everything that was wrong about modern industrial life. The many fatal collisions during the 1870s were a depressing sign of how easily corporate greed trumped considerations of human safety. Ruskin grumbled in the *Fors* of January 25, 1875, that the "ingenious British public cannot conceive of anybody's estimating danger before accidents as well as after them." Yet this was not just a question of safety. Train travel corrupted the human experience of the world. The rapid passage of trains along fixed tracks encouraged blindness to the beauty of rural landscapes and fostered slothful and wasteful habits in the passengers. On a trip from Coniston to Kirkby Lonsdale in January 1875, thoughts of the Shipton disaster weighed heavily on his mind. In Kirkby he visited one of the lookout points for Lakeland tourists, surveying the vale of the river Lune. But the place was defiled with garbage and ugly new iron seats. "Rusty and unseemly rags" lay scattered across the bank, "like the last refuse of a railroad accident, beaten down among the dead leaves." This depressing abuse of natural beauty was part and parcel of a worldview. A few months earlier, Ruskin had castigated the modern understanding of science and religion, which robbed nature of all meaning and divinity. For materialist science, even the sun was nothing more than a cosmic fluke, "a gigantic railroad accident . . ." It should come as no surprise, then, that Ruskin's mind went to trains when he felt the effects of the Storm Cloud. In August 1879, Ruskin reported in his diary that he was awakened by a "most terrific and horrible thunderstorm." The thunder rolled "incessantly, like railway luggage trains . . . the air one loathsome mass of sultry and foul fog." He read the same passage to the audience in the great lecture on the Storm Cloud.[2]

Ruskin fought these demons as well as he could. In the summer of 1875, rumors surfaced that a new railway line was to be built into the heart of the Lake District, from Windermere to Ambleside and on to Grasmere. The Kendal shoe manufacturer Robert Somervell penned a furious pamphlet against the scheme. Ruskin agreed to write the preface. Their call to arms was in fact a revival of William Wordsworth's warning against railway tourism from the 1840s. A new railroad would mar the beauty of the landscape and bring "stupid herds" of urban plebeians to the Lakes. Soon there might be no place of refuge left from industrial society: "the frenzy of avarice is daily drowning our sailors, suffocating our miners, poisoning our children, and blasting the cultivable surface of England into a treeless waste of ashes . . ." For Ruskin, the greatest threat posed by the new train line was that it would corrupt the "moral character" of the Cumbrian peasants, "whose strength and virtue yet survive to represent the body and soul of England, before her days of mechanical decrepitude and commercial dishonor." Both the land and the people had to be protected. Or, to put it more technically—Ruskin hoped to preserve the region's particular patterns of land use and fix local levels of consumption to prevent any further encroachments of modern commerce.[3]

As with so many of Ruskin's schemes, he left it to others to spell out precisely how this should be done. The key figure here was Hardwicke Rawnsley, a former student of Ruskin's who had participated in the Hinksey Dig outside Oxford. Rawnsley arrived in the Lake District shortly after the railway dispute to take up his position at Saint Margaret's Church near Ambleside. Rawnsley was idealistic and energetic but also aggressive and irritable, more of a bulldog than a sheep in Ruskin's flock of followers. One of Rawnsley's parishioners called him the "most active volcano in Europe." He quickly became involved in the railway controversy and helped found the Lake District Defence Society in 1883. That same year he established the Keswick School of Industrial Arts with his wife, Edith. He was admitted to the Guild of St. George in 1884. As Ruskin sank deeper into mental illness, Rawnsley became a leading advocate of his vision in the Lake District. Together with the social reformer Octavia Hill and the solicitor Robert Hunter, he founded the National Trust in 1895. Today, the National Trust is often seen as the custodian of a conservative inheritance industry of country houses, but for Rawnsley, the Trust was meant to carry on the radical mission of the Guild of St. George.[4]

This chapter examines the legacy of Ruskin's ethics of consumption from Rawnsley's perspective. The revival of arts and crafts was intimately connected with a project of landscape preservation and paternalist politics. Ruskin and Rawnsley treated the Lake District as a museum for traditions and mentalities that might otherwise become extinct. In this sense, Ruskin's ideals opened the door for a *vicarious* notion of sufficiency: telling poor people what they should and should not do. But the story of Rawnsley also uncovers another, more progressive aspect of Lakeland Arts and Crafts. Rawnsley's views of technology were more temperate than those of the master. He welcomed the possibility of putting steam machines to use in the service of beauty. Shortly after Ruskin's death in 1900, Rawnsley took the train from Furness to Coniston to pay homage at Ruskin's grave and see his art collection at Brantwood. This journey of the "steam dragon" into the heart of Ruskin country marked a transition away from apocalyptic fear toward a cautious appreciation for steam technology. Trains could be an integral part of pastoral landscapes.[5]

HARD PASTORAL

The Lake District is hardly a primeval wilderness. There are traces of human habitation dating back to the end of the last Ice Age. Around 3800 BC, a stone axe industry thrived in Great Langdale at the center of the region. These axes have been found across Britain. They even made their way to the Continent. In the Bronze Age, the growing population cleared the woods around peaks like the Old Man of Coniston to make room for grazing livestock and people. Distinctive local sheep breeds like the Herdwick emerged in the area, perhaps first introduced by Viking settlers. Miners dug deep into the hills. The copper mines at the Old Man may have been worked as far back as the Romans. In the woodlands along Coniston Water, charcoal was made in the Middle Ages and early modern times by coppicing the trees. Remains of "pitsteads" are scattered regularly along the shores. Coppicing preserved the trees in the landscape by culling young stems but leaving the stools and roots in the ground. The charcoal served to heat iron ore so that it could be shaped with hammers in "bloomeries." In the 1560s, German miners were encouraged to settle around Keswick to foster "modern" methods of copper mining. In the eighteenth century, the coal and tobacco trade centered in Whitehaven in the northwest part of the region experienced a brief bonanza, even temporarily eclipsing all other mercantile towns in Britain except London.[6]

Although these many industries left a mark on the landscape, they did not mar its beauty. Successive Ice Ages and a rainy climate produced the long lakes that now cut across the uplands of Cumbria. High upon the hillsides are smaller bodies of water known as tarns. Towering above them are several of the tallest peaks in England, including Scafell Pike, Skiddaw, and Helvellyn. Yet their scale is more human than the majestic and forbidding vistas of the Alps and the Rocky Mountains. For the most part, the drama of this landscape is accessible to reasonably fit walkers. It is possible to traverse the entire region in a few days' ramble, with a pint of ale and a soft bed as a reward each night. Yet there are also places of unexpected wildness and pockets of high biodiversity amid the patchwork of sheep farms and quarries. The summits become treacherous in poor weather. There are cliffs steep enough to injure or kill. Even at the height of the summer tourist season, there are times of solitude on the peaks, especially in the early morning or late evening. Perhaps the greatest number of rare species can be found in hidden gorges along fast-flowing streams where sheep cannot graze.[7]

The Lakes became a popular destination for British tourists in the late eighteenth century. William Wordsworth gave the region an enduring poetic expression in the *Lyrical Ballads* of 1798 and 1800. Wordsworth was fascinated with the rugged character of native farmers and shepherds. He thought that the rough hills and poor soils of the region ennobled the mind by fostering habits of independence, industry, and self-sufficiency. This was not the easy world of leisure idealized in pastoral poetry and painting, but a moral code steeped in constant solitary work, "[w]ith a few sheep, with rocks and stones, and kites, that overhead are sailing in the sky." Wordsworth's poem "Michael" tells the story of an aging shepherd who sends his son away to pay off a debt in order to secure the patrimony of the farm. He loses both when his son is corrupted by the vices of the city. Yet Wordsworth's sympathies remain with the shepherd. Michael's only error is that he loves the farm "even more than his own Blood." Indeed, his character seems inseparable from "These fields, these hills which were his living being." The landscape, that is, "[t]he common air, the hills . . . impress'd so many incidents upon his mind, of hardship, skill, or courage, joy or fear." For Wordsworth, the character of the Lakeland farmers was not just an interesting poetic subject but also the foundation for a stable political and social order. In a famous letter to the Whig politician Charles Edward Fox, Wordsworth explained that these "small independent proprietors of land here called statesmen" were the bulwark of liberty and social harmony. Their "daily labour on their own little properties" offered a role model for the poor everywhere in Britain. In a time of high bread prices and near famine, the government should support this "spirit of independence" rather than trap the underclass in a system of subordination based on "workhouses, Houses of Industry, and . . . Soupshops."[8]

"Michael" was one of Ruskin's favorite poems. He saw Wordsworth's description of the shepherd as an ideal example of the "Border peasantry of Scotland and England." It is not difficult to see why. The shepherd and his wife live a simple life of "endless industry." While Michael works in the hills, Isabel spins her own flax and wool at home. Among their few possessions is an old lamp—"an aged utensil"—which shines in the window of their cottage every night, as constant as the "Evening star." Their diet consists of "pottage and skimm'd milk . . . with oaten cakes and . . . plain homemade cheese." But they are content in their poverty. "We have enough," Michael tells Isabel. For Ruskin, this was an accurate portrait of Lakeland life even in his own day. The Westmoreland farmers, he insisted, were "hitherto a scarcely injured race" that represented the true "body and soul of England." It was this kind of peasant virtue that the new railway threatened to undo. An influx of visitors of the wrong sort from the "suburbs of our manufacturing towns" would destroy the high character of the natives. In a letter from 1884 to the *Manchester City News*, Ruskin poked dark fun at the boosterism of the Ambleside hotel lobby. A railway could improve the commercial prospects of the town immeasurably. He joked that there would be "magnificent establishments in millinery and 'nouveautés,'" an esplanade around Rydal and Grasmere Water, a "Lift to the top of Helvellyn, and a Refreshment Room on the summit." The Vale of St. John would be "laid out in a succession of tennis grounds" and would also boast a Casino "decorated in the ultimate exquisiteness of Parisian taste." In contrast, the virtues of the "statesman farmer" required isolation from modern commerce and transportation to thrive. The harsh but beautiful environment fostered a taste for the simple life. Hence, it was necessary to protect the Lakeland landscape and people at the same time. "It would be hard to find a more powerful plea than this," Robert Somervell wrote, "for the preservation of the Lake District." The campaign against the railway was in effect a kind of moral experiment that sought to make out of Lakeland a great outdoor museum with live human subjects in the habitat.[9]

Effective preservation required political action. Ruskin played the role of the prophet with his fiery denunciations, but proved unwilling to engage in the practical side of the work. Instead, he cultivated local allies who could undertake the task of organizing voluntary associations and petitioning Parliament. The pamphlet against the railways became the beginning of a lively friendship with Robert Somervell. Ruskin was so impressed with Somervell's zeal that he recruited him into the Guild of St. George, where he served as legal counsel. Ruskin helped Somervell's campaign not only by writing the preface to the manifesto but also by giving him access to his enormous network of correspondents. He advised Somervell on who might be most open to their message: "It is especially a petition in which women's voices should be of weight and then of youths whose character may be chiefly formed among noble national scenery." In the end, neither the Ambleside railway nor any other line into the center of the Lake District was ever built. The boosters of the railway seem to have abandoned it for fear that it would be unprofitable rather than out of respect for Somervell's campaign.[10]

However, this was only the beginning of the struggle. The next great fight started in the summer of 1877, when the Manchester City Council voted to establish a reservoir in the Lake District. Like other northern communities, Manchester had experienced explosive population growth during the Industrial Revolution. The city quadrupled in population in the first fifty years of the nineteenth century. Massive construction accompanied this boom. Yet the water supply lagged. In the midst of so much manufacturing industry, clean water was becoming a scarce resource. At first, the city favored a corporate solution by way of the private Manchester and Salford Waterworks Company. But the approach proved ineffective. Short-term profit maximization did not mix well with the goal of providing clean water for the public good. Changing tack, the new municipal government of Manchester established a reservoir in Longdendale Valley, west of the city. But they failed to anticipate the growth of the city. The new reservoir proved inadequate for demand almost as soon as it opened. Instead, the city government turned east to the Lake District. Here was an abundance of clean water that could be piped down to the city. After analyzing the topography of the region, the surveyors and engineers on the waterworks committee settled on Lake Thirlmere, northwest of Wordsworth's Grasmere. It had the right volume of water and proper elevation. The major challenge was to blast a tunnel through the watershed at Dunmail Raise so the water could flow south and east to Manchester.[11]

News of the scheme caused outrage among the people who had opposed the Ambleside railway. *The Times* printed a furious letter by Ruskin on October 20, 1877. Why was the government of Manchester "plotting . . . to steal, and sell for a profit, the waters of Thirlmere and the clouds of Helvellyn"? The Thirlmere Defence Association was established the same year, with sixty people in attendance at the first meeting. Somervell was appointed the secretary of the society. Along with the *Pall Mall Gazette*, he tried to turn the arithmetic of water use against the developers, claiming that Manchester did not really *need* the water, since it was actually selling more of its water than it was consuming: "[E]leven million gallons of water are daily sold for manufacturing purposes and to outsiders, against six million sold for sanitary purposes and domestic use." This was essentially an argument against the increase of manufacturing capacity. But for the majority of the critics, another tactic held greater appeal. Although Thirlmere did not share the literary reputation of Grasmere or the other, more famous lakes, they stressed that it had unique qualities of its own. J. Clifton Ward, for instance, wrote to the *Daily News* in 1877 to emphasize the rarity of the valley, noting that England "has but a few rock-embosomed lakes, but each [is] unrivalled in its way." He appealed to the higher sentiments of the reading public, pleading that "Nature in her purest aspect" and "wildest moods" ought to be defended. Yet, in the same breath, he had to admit that "sentiment" could do little against "purely utilitarian measures" without wider support. As *the Standard* reported, there was "no opposition to the scheme" from the people living at Thirlmere, "who only number two hundred. The whole opposition comes from without." This was precisely the problem with which Rawnsley and his peers grappled during the campaign. Many locals had little education; fewer still read poetry. It was far from certain that they would put the preservation of the landscape ahead of other pressing issues of livelihood and profit. Some of them might be easily swayed by developers, who promised great sums of cash. The best strategy, then, was to claim Thirlmere as a national property for the British public at large, a place of unblemished beauty in which to experience "direct contact with nature." Promoters of the waterworks scheme heard the complaints of educated protesters, and tried to meet their accusations head on. They generally argued that, far from destroying the beauty of Thirlmere, they would actually improve it while also enabling better access to the region. New wide, smooth roads, they insisted, would allow more tourists to see Thirlmere than ever before; they would also be a boon to tradesmen in the area who might profit from increased traffic. But the critics scoffed at this notion. John Hammon Fowler, for example, predicted the destruction of all opportunities for employment when the dammed water drowned existing local communities. It was a "folly" to think people would come to Thirlmere once the waterworks had been finished, wider roads or no. It was, after all, difficult to reach and possessed a rugged beauty less appealing to average vacationers.[12]

In the end, utilitarian arguments prevailed, and the scheme went ahead. The Thirlmere reservoir was formally opened on October 12, 1894. Several small properties along the lakeside were flooded. The little village of Wythburn was evacuated. Only its church and a tavern remained above water. But the destruction of the landscape at Thirlmere was not the decisive triumph for the commercial and industrial interests in the Lake District that some had feared. On the contrary, the next year saw the establishment of the National Trust, an organization firmly committed to regulating land use and limiting commercial development, which would eventually acquire control over more than a hundred thousand acres of scenic land in the region. The ghost of Wordsworth thus prevailed over the boosters of growth. A key figure in this achievement was Hardwicke Rawnsley.[13]

BONBON DISHES AND SPINNING STOOLS

In 1877 *the Bristol Mercury and Daily Post* reported on a "Fashionable Wedding [n]ear Ambleside." The bridegroom and vicar, Hardwicke Rawnsley, and his bride, Edith Fletcher, gathered with some fifty guests inside Brathay Church at 10:30 a.m. Edith was "attired in a dress of soft cream drawn silk," and against the cold February air she wore an "over-dress" too, "with full train . . . of brocaded silk," "epaulettes," and a "high collar"—all "trimmed with white fur." In her hair were "natural orange blossoms. There were no fewer than seven bridesmaids, each wearing "pale blue cachemire and silk" and stylish wide-brimmed "Gainsborough hats" with lace and ostrich feathers. The bridegroom had given each of the bridesmaids monogrammed silver lockets to wear. Carriages took the guests from the church to a formal breakfast. At noon, the bride changed into a "travelling costume of sage green" and the couple was escorted on their way across a carpet of "flowers strewn at the entrance by the assembled servants." The couple then bade farewell to their guests and took the afternoon train to the south. For such a small village in the north of England, the entire affair was very fashionable indeed.[14]

But this display of wealth and taste was not entirely conventional. Nor was it designed to cater to women's desires alone. The bridegroom attended carefully to the details of the event. He wrote all the hymns for the ceremony, and even began the marriage service himself. He also insisted that in place of the Homily, a speech was given by Edward Thring, his old headmaster at Uppingham, a public school. Thring was famous for his egalitarian views regarding working-class education and his efforts to broaden the academic curriculum. He admired Ruskin and believed that a student without talent in language or mathematics might yet "be clever with his hands." He encouraged the study not just of classics, but of music, art, and even woodworking. At Uppingham, he converted the old archdeacon's room into a carpentry workshop for the students. It was Thring who had first brought Rawnsley to the Lake District on a school holiday.[15]

Under the influence of his headmaster, Rawnsley had gone up to Oxford in 1870, where he studied with Ruskin. He participated in the road construction at Hinksey village and was introduced to the urban reformer Octavia Hill. Ruskin inspired in him a radical view of pastoral work. Why not add to each church and mission room "a parish workshop, where the blacksmith and the village carpenter shall of a winter evening teach all the children . . . the nature of iron and wood, and the use of their eyes and hands"? This idea of combining religious work with Ruskin's ideal of handicrafts profoundly shaped Rawnsley's career. After volunteering in a London hostel for the destitute, he suffered a nervous breakdown and went to Windermere to convalesce with friends. Ordained as a priest in 1877, Rawnsley gained an appointment to St. Margaret's Church in Wray, outside Ambleside. Rawnsley thus arrived in the region just as the fight over Thirlmere was beginning to stir. He was close enough to Brantwood that he could renew his acquaintance with Ruskin on regular walks "over the Hawkshead Hill." The Rawnsleys soon became preoccupied with the idea of fostering the Arts and Crafts in their own parish. They began to offer classes in wood carving at Wray in 1880. The response was evidently enthusiastic. Pupils came all the way from Grasmere and Ambleside. In order to keep up with demand, they had to employ a "lady teacher" from South Kensington.[16]

When Rawnsley was appointed vicar of Crosthwaite at Keswick, these ambitions to revive Lakeland crafts expanded in scale. The couple set up the Keswick School of Industrial Arts in 1883, the same year that Albert Fleming founded the Langdale Linen Industry. Eventually, the range of the school came to be far wider than Fleming's scheme. It trained workers in the production of handcrafted goods made from wood and silver, copper, and brass, not to mention linen. A "List of Articles Made at the Keswick School of Industrial Arts" includes all the typical home goods, such as silver tea services, napkin rings, porringers, and candlesticks. The inventory also incorporated specialty items like silver-plated copper "Fern Pots" and "Muffin Dishes" as well as silver spoons done in various patterns. In copper and brass, one could find items such as cigar caskets and coal boxes, boxes for biscuits or gloves, or even "Paper Knives, Pen Trays, [and] Pen Wipers." In silver, they made "Bonbon Dishes, Eau de Cologne Bottle Covers, [and] salt cellars," while for jewelry they listed buttons and hat pins. Wood carving was popular, with work including "Bread platters, glove boxes, [and] lamp stands."[17]

Caught in a paradox of their own making, the Rawnsleys tried to turn Ruskin's model of industry into a thriving business. A surviving photograph of the KSIA showroom gleams and glints with silver platters and bowls and other objects not unlike those seen in the great displays of urban stores. Wooden objects on sale included "potato bowls" and "spinning stools." The latter were sold as part of a strategy to encourage public interest in local traditions and to demonstrate the people's ancient connection to the land through the wool and flax that had been historically produced there. With an eye to his own profession, Rawnsley also targeted ecclesiastical needs. Churches could be counted on to purchase objects for display and ritual, such as "Alms Dishes[,] Altar Book Rests, Hymn Boards, [and] Chalices." The repoussé metalwork for sale sported designs inspired by local archaeological digs (often under the direction of W. G. Collingwood), whether Celtic or Etruscan or the "St. Kentigern's" style, invented specifically for the KSIA, as Saint Kentigern was the patron saint of Crosthwaite Church. Their relatively plain style would have stood out against the ornate patterns that had become widely available thanks to new mechanical production technologies. Local interest was also emphasized in trays that came in "Skiddaw" form, as opposed to merely "oval" or "square cornered." Although Ruskin praised the merging of use and beauty, the KSIA was not above selling purely decorative silver and enamel brooches and pendants as well.[18]

It was a modestly profitable enterprise. Initially, they struggled just to break even. Rawnsley calculated that they had expended £181 but "produced work which we estimated to be worth £118." Fortunately, the situation soon

FIGURE 4.1 Hot-water jugs designed by Harold Stabler and executed by Thomas Spark, 1899. Both were associated with the KSIA. See Esther Wood, "The Home Arts and Industries Exhibition," *The International Studio*, Charles Holme, ed., August 1899, 104.

improved; Rawnsley never intended the KSIA to be a charity industry. It is likely that Fleming was called on for advice. During the third season, "we found ourselves with our expenditure doubled but our sales had doubled also." Rawnsley estimated their work to be worth £1,700 a year in 1899. He proudly described the industry as "entirely self-supporting"—something Fleming had been keen on showing as well. Throughout all these years of hard work, Rawnsley always underscored the idealistic origin of the industry. John Ruskin was "the spirit that had made the whole venture possible" in the first place.[19]

Though Fleming and Twelves had made the initial foray into the production and sale of handicrafts in the Lake District, it was Rawnsley's KSIA that held the first great handicraft exhibition in 1885. This soon became the preferred venue for local crafts. At these exhibitions, Ruskinians from all quarters of the Lake District came together. Workers from the Langdale Linen Industry, artisans of Arthur Simpson's workshop, as well as Annie Garnett's Bowness spinners and weavers converged to stage what could be described as a clever, pleasant, multifaceted protest against modern industrialism. From 1885 all the way up to 1920, local exhibitions offered a crucial market for "revival" crafts. They also presented an opportunity to build solidarity within the movement. There were a large number of "inaugural addresses, lectures or closing speeches," which reached a substantial and sympathetic audience all over the Lake District. The 1891 exhibition at Kendal proved particularly influential. Rawnsley gave a lecture that disapprovingly detailed the "'hoveldom and haddledom' of manufacturing towns," that is, the material squalor and bad morals of the urban working class. Lord Carlisle in turn recognized Ruskin's guiding spirit in his speech. But despite the Kendal exhibition being as successful as anyone could possibly have hoped, its message failed to reach beyond the regional community it aimed to protect. The event received no attention in any of the major national newspapers or journals. This was lamentable but also predictable. After all, many of the Ruskinians aimed to make the region self-sufficient and insulated from the trends of the greater nation.[20]

TENDING THE FLOCK

In the same year that the Rawnsleys founded the Keswick School of Industrial Arts, the railway controversy erupted again in the Lake District. At the initiative of the local landowner H. C. Marshall, the proprietors of the slate quarries at Honister Pass wanted to build a railway along the western shore of Derwentwater. Rawnsley got wind of the scheme and brought it to national attention with a letter to the *Standard* in February 1883. He warned the public that the new railway would spoil forever the famous view across the lake from Keswick: "Let the slate train once roar," and it will "damage irretrievably the health, rest and pleasure" of Englishmen who came to visit the region. Rawnsley's letter reignited the debate between advocates of economic development and preservation. In contrast with Thirlmere, the landscape at Derwentwater had a richer store of literary associations that the preservationists eagerly seized on to make their case. A Defence Fund was established that quickly raised the sum of £402. Ruskin endorsed Rawnsley's petition against the new railway, but was pessimistic about the outcome: "You will soon have a Cook's tourist railway up Scawfell and another up Helvellyn and another up Skiddaw, and then a connecting line all round." Despite Ruskin's pessimistic forecast, the preservationist side prevailed. To ensure that there could be no new initiatives of this sort, Rawnsley and his allies founded the Lake District Defence Society in the aftermath of the crisis. Speaking to the Wordsworth Society at Westminster in the spring of 1883, Rawnsley observed:

> Our only chance of keeping Lakeland inviolate is to be on the watch with a powerful national, one might say international committee . . . with a backing of Members of parliament to help us at Westminster, and a considerable sum of money behind us for expenses.

Rawnsley and Albert Fleming became the first joint secretaries of the society, together with Fleming's friend W. H. Hills, a London bookseller. Most of the 400 members of the society were outsiders rather than locals; "fewer than 10% were from Cumberland or Westmoreland." More than half hailed from greater London or Manchester. Leading members included Ruskin, Tennyson, William Morris, and the Duke of Westminster. In short, the society was supported by a prominent faction of the cultural elite of England, an outside group without an economic stake in the region.[21]

Over the next few years, Rawnsley spent much of his energy on other, related issues of preservation. He joined the opponents of the Thirlmere reservoir, campaigned for public footpaths, and sought to block the introduction of telephone lines, among other things. Most famously, in 1895 he joined his old friend Octavia Hill and the solicitor Sir Robert Hunter in founding the National Trust for Places of Historic Interest and Natural Beauty. For Rawnsley, the purpose of the Trust was very much a continuation of Ruskin's aims. In a letter to Edith Hope Scott, he drew an explicit parallel between the National Trust and Ruskin's Guild of St. George. To preserve "Beautiful and Historical England to posterity," he noted, "we ought to have a National Trust Circle, wherever St. George's knights are." In practice, however, the priorities of the Trust were rather more limited than Ruskin's organization. The guild's radical aim of reviving cottage industry never became central to the National Trust (although Rawnsley certainly promoted this other economic goal with the Keswick School of Industrial Arts). Instead, the main mission of the Trust was to preserve "lands and tenements (including buildings) of beauty or historical interest." Around the turn of the century, the organization began to acquire land in the Lake District. The first site was the memorial to Ruskin on Friar's Crag, followed by 106 acres at Brandlehow Park, both on the shores of Derwentwater. Twenty years later, the Trust had become a major landowner in the region, holding over 1,387 acres. After World War Two, this stake grew to 122,000 acres.[22]

For Rawnsley, the preservation of the landscape went hand in hand with the protection of the people in it. He was convinced, like Wordsworth and Ruskin, that the native people were the descendants of an ancient race devoted to liberty and a rustic way of life. In the founding documents of the Lake District Defence Society from 1883, he spoke of the "Cumbrian and Westmoreland peasant" who "for all the attempts to spoil him, for all the tourist prices and presents, is as yet a character unspoiled"; "these dalesmen," he added, "are made of such rare peasant stuff that it is worth preserving. Slow of song, brief of speech, but sure of word as they are." He had said something similar in a presentation before the Wordsworth Society in 1882 when he noted that these people led a "severely simple life," yet it was one "with 'joy in wildest commonalty spread.'" Ruskin too had praised the old-fashioned morals of locals, observing that he "could take his tradesmen's word for a thousand pounds, and need never latch his garden gate, nor fear molestation in wood or on moor, for his girl guests." Both men thought these virtues were now imperiled. New railway lines and the growing importance of outside money threatened to undermine the unique character of the local population. Worse, the Lakeland people could not be trusted to resist these temptations. They were "not the safest guardians of their lovely homes," but needed enlightened leadership to protect the country from corrupting influences. Much of Rawnsley's life as a vicar and preservationist was devoted to this task.[23]

As one might suspect, such a paternalist mission to protect local people from factories and mass consumption was full of pitfalls. Albert Fleming, for all his attempts to find meaningful employment for poor women, seems to have been unable to acknowledge fully Marian Twelves's contributions; and if he could not appreciate her, an educated and articulate housekeeper from the city, how much did he really understand the poor women whom he employed in the Langdale Linen Industry? To his credit, Rawnsley was aware of the gap between the high ideals of preservation and the outlook of the farming families that he wanted to help and protect. The lure of better jobs and higher pay was powerful. Why should poor people care whether factories and railways were built in the area? He knew he needed to win the hearts of native Lakelanders, and seems to have made earnest efforts to understand them better. His studies of Wordsworth's life in the Lake District made clear just how difficult it was for "off-comers" to gain the respect and trust of local people. As early as 1882, Rawnsley endeavored to highlight the legacy of Wordsworth in local tradition. In "Reminiscences of Wordsworth among the Peasantry of Westmoreland," Rawnsley interviewed several former employees of the poet, faithfully recording their memories in transcriptions of the native dialect. In the process, he discovered a strange gap between the poet's vision of the Lake District and the perceptions of local people. Like Wordsworth, Rawnsley thought the Lakeland farmers were a race apart—"Nature's gentlefolk"—but their noble character was now under threat. "Strangers with their gifts of gold" were corrupting the poor. The farmers had to be reacquainted with the "immortality of lofty tradition" that Wordsworth had celebrated in his poetry. Yet Rawnsley's interviews also made clear that Wordsworth had venerated local people without ever really getting to know them. How could you convince people to return to their true, noble spirit when they had no reason to listen to you?[24]

Rawnsley carefully transcribed the speech of his informants, obviously fascinated by their native dialect (he noted a number of elocutionary oddities—for instance, how "quite" was pronounced "white") and what he took to be their obstinate, slow ways: "The Cumberland mind is not inventive, nor swift to anticipate the answer you wish." His "first witness," he explained, was a former employee of the Wordsworths at Rydal Mount who possessed a "practical and unimaginative mind." When the woman described Wordsworth as an "ugly-faäced man, and a meän liver," he assured his readers this merely meant "that he was a man of marked features, and led a very simple life in matters of food and raiment." One man, who used to have the dissolute son of Samuel Taylor Coleridge as a lodger, was of the surprising opinion that Hartley Coleridge contributed "t' best part o' his poems for him, sae t' sayin' is," and that while Wordsworth was "a cliverish man," he nevertheless "wasn't set much on by nin on us." Another man, a builder, suggested that Dorothy Wordsworth "did as much of his potry as he did." The local people, in other words, had seen Wordsworth face to face, sized him up, and found him singularly unimpressive. It was a curious world where "white" did not mean white, questions were followed by non sequiturs, and Wordsworth was not considered a great poet. Doubtless, Rawnsley was both amused and appalled by these sentiments. Yet he diligently sought the cause of their disregard for the poetic genius who had aimed to protect their very own Lakeland countryside. A man described as "a splendid type of the real Westmoreland gentleman labourer" indicated that, as Rawnsley put it, Wordsworth was "shy and retired, and not one who mixed freely or talked much with them." "He was not a man as fwoaks could crack wi'," another said. A former butcher's delivery boy noted, "But as fur Mister Wudsworth, he'd pass you, same as if yan was nobbut a stean. He niver cared for childer . . . he niver oncst said owt." Rawnsley showed that these working men and women were far removed from the intellectual and spiritual sensitivity of the poet, while also indicating how they were emotionally perceptive. If they were not imaginative or sharp, they had a strong sense of family and community. Wordsworth had missed the opportunity to connect with them and convey the subtle, lasting value of his poems and life choices. The people described Wordsworth and his wife as plain, if not downright ugly, overlooking the more important fact that they had beautiful souls and were attracted to these spiritual qualities in each other. Nor did they understand why Wordsworth spent so much time writing, as Rawnsley put it, "for sheer love, and not for money." To Rawnsley, their innocence was a sign of latent nobility, but also a fatal flaw. They needed help and guidance to resist the spell of materialistic values.[25]

Typical of Rawnsley, he combined acute insight with obtuse condescension. He imagined, much as Wordsworth did, that the peasants felt "joy" in their difficult rural lives, but he had a myopic paternalist notion of the simpleminded happy poor, assuming that they were devoid of desires or aspirations beyond sheep farming. Wordsworth's "seclusion, and the distance he seems to have kept from them and their cottage homes," he found "not a little interesting." This distance resulted, he felt, in a blindness to the people's negative traits. Everything suggested that "the poet lived so separate and apart from them, so seldom entered the 'huts where poor men lie,' . . . that he was enabled . . . to forget, quite honestly perhaps, the faults of the people among whom he lived." Rawnsley was oblivious to his own tenuous connection to the people, even as he slowly uncovered Wordsworth's imperfect relationship with them. But he recognized the irony of the people's love of Hartley Coleridge, the poet and son of Samuel Taylor Coleridge. Hartley had succumbed to alcoholism as a young man, which destroyed his budding career at Oxford. He retreated to the Lake District, where he was cared for by a local couple. Yet Hartley became a favorite with adults and children alike, and wrote a small number of poems that they adored. Hartley was a man of the people, whereas Wordsworth stood apart—despite the fact that he lived simply and prized the vernacular voice in poetry. Lakelanders were not impressed by Wordsworth's abstemious diet and philosophical bent. They preferred Hartley's social ease and common habits, however ill he might have been. Hartley fit into their way of life, while Wordsworth was "niver a frequenter of public-houses." Nor, it should be added, was Rawnsley.[26]

Rawnsley seems to have found inspiration, and a vivid counterexample, in Ruskin's surprisingly congenial relation to local inhabitants. "It may be true," he wrote in 1902, "that Ruskin stood in nearer relation to the peasants than the poet. Wordsworth, as the tradition in the dales still goes, 'was not a vara conversable man at best o' times.'" But "with Ruskin," he said, "what he seemed most to care about was to go to the cottage or to the workshop, and be made by his tender approachableness one of the family." Ruskin was far more comfortable interacting with the locals. He devoted considerable time and energy to the development of the village school and activities for local children, all of which won him favor, though he was always known as "the professor," a man apart from everyone around him. Perhaps the difference in how Wordsworth and Ruskin were perceived came down to their material circumstances. At one point, Rawnsley suggested that Wordsworth's lifestyle was *too* simple, and even severe. He mused that Wordsworth might not have approved of the elaborate ornamentation involved in the making of handwoven linens according to Ruskin's taste: "he would have wondered what fine ladies could be found here-about — from princesses downward — to deck themselves in rainbow hues, and wear Greek and Sicilian laces wrought in Cumberland." In contrast, Ruskin seems to have hit upon an appealing balance between the simple life and self-adornment, ethical consumption and aesthetic taste. He never felt that one had to preclude the other.[27]

As a vicar, however, Rawnsley had another powerful social strategy to try as well. He believed that religion and ritual offered an alternative repertoire of communication and persuasion. He was keen to infuse into old Lakeland traditions the gospel of Ruskin's ethics of consumption. Perhaps the best example of this was the Rushbearing ceremony held at St. Oswald's Church in Grasmere (and elsewhere), usually at the end of July or beginning of August. In an annual sermon, Rawnsley preached the virtues of the simple life to farmers and visitors alike, a practice he continued for many years, until the Great War. He sought to persuade his listeners, many of whom were Grasmere farmers, that keeping up local traditions was a moral necessity. Rawnsley fondly addressed the churchgoers as "sons of the Vikings who dwell in the homes of the Britons of old, in this Grasmere Vale." If he could convince people that their ancestors had always revered and cared for the land, it would seem more natural for them to do so in the present. This was "the power of a simple thanksgiving service to bind century to century and generation to generation." During the Rushbearing ceremony, reeds were carried from the surrounding woods or fields to be strewn across the church's floor. It was an ancient custom designed to warm and dry the churches, which in earlier times had floors of bare earth. The reeds rendered the cold, dank churches fragrant and cozy, a place fit for gathering. Rawnsley impressed on his listeners the centrality of vegetation and fertility as well as the solemnity of community: "Let us also not be unmindful," he said, "of the generations of Grasmere people . . . that they remembered the days that were past . . . as they strewed the earthen gloom of their village church with reeds." It was a compelling way of extending the community both forward and backward in time, bringing out the antiquity and strength of local regard for nature. The association stretched as far back as the Roman festival of Floralia. Rawnsley stressed that this lively, physical, religious activity "was not a thing fit only of sad souls . . . but a thing for hours of homely merriment and innocent pleasure," for children as well as adults. His vision was fixed on the future, and he knew that success depended on persuading young people in the parish to continue the ritual. The carrying of rushes, like Fleming's spinning project, was a tradition palpably connected to the earth—an ancient tradition he believed to be on the verge of extinction.[28]

The Rushbearing ceremony was not uniquely connected to Rawnsley, but it is hard not to feel that his version of it was decidedly tinged with Ruskinian ideals. At times Rawnsley turned his Rushbearing sermons into an open critique of modern forms of consumption. He often told the story of Theobald, who "felt called to forsake all that he had for Christ and go off with a companion into the forest near to Reims." The life Theobald chose was one of "simple abnegation and self-sacrifice for Christ's sake." In the modern world, it was easy to forget the value of material simplicity: "We in this age of wealth and comfort and luxury are more and more surprised that anyone for Christ's sake will prefer to be poor." But, again stressing the pedigree of Lakeland people, Rawnsley set them apart from the moneygrubbing, soft-handed city dwellers. "Now many of us are simple laboring men. We follow sheep to the fell; we take our horses to the great Fair at Brough Hill. We sow and reap; we cut & carry the grass, are farm labourers, blacksmiths, or journeyman joiners," he said, using an inclusive "we." He wanted to make the lower ranks of the parish feel connected to him as well as to the land, so they might be content with whatever variety of simple work they happened to do. In an undated "Harvest Sermon" on "The Dignity of Manual Labour," he contrasted "skilled labor . . . crowded into towns" and the "simple labor of the fields." He emphasized that a man never stands "in nearer relation to the creator . . . than when he tills." Farmers held a higher rank in the spiritual hierarchy than "the producer of comforts and luxury."[29]

FIGHTING THE STORM CLOUD

On October 12, 1894, when the decades-long Thirlmere Reservoir scheme was finally completed, Hardwicke Rawnsley surprised many by presiding over the opening-day celebrations. The great champion of the Cumberland peasantry began the ceremony with a prayer to bless the new dam and the waters that now enveloped a small village. He also contributed four commemorative sonnets specially written for the occasion. It was a serious coup for the Manchester Corporation to secure his presence. One reporter concluded, "When a poet—for Canon Rawnsley is himself a poet . . . declares that the engineer has accomplished his mighty purpose, without spoiling the district, there is nothing more to be said." The opponents to the scheme, the journalist suggested, had been entirely misguided, for everywhere locals cheered the celebratory procession on, looking upon them not as "Goths and Vandals . . . but as benefactors." This was of course a very selective view of the situation. Why had Rawnsley swung around from the preservationist position to cheer publicly the construction of the reservoir? His dismayed erstwhile friends and allies guessed that he had somehow been corrupted by promises of compensation from the Manchester lobby. W. D. Hills thought that a "party of innkeepers" had put pressure on Rawnsley to acquiesce to the building of a new road past Thirlmere. But this is not the whole story.[30]

The *Manchester Times* journalist recounted the procession of the Manchester delegation and local dignitaries from the Windermere train station to Thirlmere. The reporter was careful to point out that Windermere was the "noblest of the English lakes," in sharp contrast to Thirlmere, whose "towering mountains" and "giant shapes" were "weird" and "uncanny." As the party approached Thirlmere by way of Dunmail Raise, they entered a "bleak" and forbidding valley. By building roads and damming the lake at Thirlmere, the engineers had turned what appeared to be a "straggling river" into "a far nobler stretch of water." The altered Thirlmere was now almost as beautiful as Windermere, the reporter implied, and as accessible. But Rawnsley's four sonnets written for the occasion did not offer unequivocal praise. The damage done to the valley was "too deep for any time to heal." Although the area was not as well known or loved as others in the Lake District, it had a rich literary history of its own, which was now far more difficult to appreciate. Rawnsley and his wife, Edith, had fought to save the Rock of Names at the south end of the lake from destruction. This stone was inscribed with the initials of Wordsworth and other figures from the Romantic era, according to local lore. When the engineers destroyed it, the Rawnsleys carried the remnants up the hill to form a cairn by the new road. Whereas Rawnsley's first commemorative sonnet praised the engineer who "had gauged the future, felt the stress/Of that great city's toil and thirst and strife," another poem, "To the Workers," ended with a deep sense of loss:

> [amid] far bleat, and raven's call, and phantom horn
> Of Viking chief upon his ancient hill
> I caught the monotone of pick and drill . . .

But the note of mourning proved too subtle for some. The reporter for the *Manchester Times* printed two of Rawnsley's four poems without the slightest acknowledgment of any nostalgia for what had been lost. Instead, Rawnsley was described simply as "the author of one or two happy sonnets on the Thirlmere scheme."[31]

In truth, Rawnsley's reluctant embrace of the new dam was born of a deeper recognition that technology might have a place to play in preservationism. Like many other followers of Ruskin, Rawnsley had read the apocalyptic lectures on "The Storm Cloud of the Nineteenth Century." In the article "Sunlight or Smoke?" for the *Contemporary Review*, published in 1890, Rawnsley echoed some of Ruskin's premonitions. But although he shared Ruskin's fears for the devastation of nature and mankind along with it, he also toyed with the possibility of a technical fix to the problem. The report described Rawnsley's visit to the colliery in the town of Farnworth in Lancashire. His first order of business there was to assess just how bad conditions were for the people who lived and worked there. He found that "the heavens were black with smoke," and noticed with horror how the "sulphurized air" diminished the lives of the colliers, who "in their clogs clattered by, grim and grimy." The scene was comparable only to the hellish landscape of Dante's *Inferno*. Nor was the desecration limited to human life: in front of the remains of a "grand house," "smoke-blighted trees" "turned withered arms upward" "like souls in pain." Beneath a "dark, sooty hill . . . crept . . . a thing that only in Lancashire could be called a river." The waters were "poisonous with the discharge" of local sewage tanks, and were "black with the refuse waters of mines and chemical works for miles." It is hard to say whether the reader was supposed to sympathize more with the colliers or their grim landscape. Yet in the midst of the slag heaps and collieries of Farnworth was a new industrial development, which promised to "solve the smoke problem for England." This "smokeless chimney" was in fact the reason for Rawnsley's visit, a breakthrough in manufacturing technology that "for these past twelve years, had been pleading for light and wholesome sunny air."[32]

Rawnsley took great care to describe just how the chimney worked after a tour of the plant with his friend the proprietor. The crucial innovation was to use "mechanical stokers" when burning coal to produce steam. By feeding fuel slowly but regularly into the furnace, impurities were removed and smoke reduced. The new technology had been introduced by Lancashire coal-mine owner Herbert Fletcher, who was a member of the newly formed Committee for Testing Smoke-Preventing Appliances. Rawnsley noted that the system involved a mechanism for making use of "coal refuse"—it was not only smokeless and clean, but efficient and able to take advantage of "waste." With the optimism of the amateur trying to grasp a science he respected but did not fully comprehend, he noted that "none of the deadly carbon monoxide" or "carbonic acid" was present, and that "it would be contrary to first principles in chemistry that he should find it." A common objection to smoke-abatement technology among Rawnsley's contemporaries was that it reduced smoke but increased instead emissions of lethal carbonic acid (carbon dioxide) and carbon monoxide. At the Farnworth Bridge colliery, chemical analysis showed that the smokeless gases emitted contained 80 percent nitrogen, 14 percent carbon dioxide, 4 percent oxygen, and 2 percent sulfur. They were dispersed so rapidly, Rawnsley claimed, that no harm was done to any living beings. He "left Farnworth devoutly thankful" to this friend who had been working hard to get the word out about this new technology, so that those in the industry could "see for themselves how easily and at how small a cost and how effectively the smoke demon could be combated." Yet he felt certain that he had seen "the battle of the future," which he felt would arise between the faster, older methods of stoking, and these newer, *slower, cleaner methods*.[33]

Rawnsley left Farnworth for Bolton, where he toured six other factories. Most were attempting some level of smoke abatement using new technologies. Thanks to such "public spirit," the pollution was greatly reduced. This sacrifice, he emphasized, was "[n]ot a great outlay," considering that the new method allowed for a cheaper kind of fuel made up of "waste" materials. But so far, inventions like this one failed to be widely effective, due to confusion surrounding conflicting standards in smoke-abatement laws. What was illegal in Manchester, for instance, was still acceptable in Bolton. He found evidence of resistance after visiting "one of the largest flannel factories in Lancashire," which had returned to the older, more polluting method of stoking. It came down to the speed of production: "It pays some firms better to work with quick firing and a better quality of coal . . . than to go in for slower furnaces and more land or larger boilers and chimney." Certain manufacturers—perhaps most—were "devoid of a public conscience." At the same time, the penalty for abusing new public health laws was not severe enough to force anyone to truly change their ways. As Rawnsley made his way home that day, an advertisement for "Sunlight Soap" caught his eye. When he brought it to the attention of a clerk on the train, the latter commented, "'that's the only sunlight we chaps gets [*sic*] in Lancashire.'" His article ended with a prayer in favor of smoke mitigation:

Let the Furnace-owners realize that smoke-prevention is their duty.
Let the workmen understand that smoke does not mean work . . .
Let electors feel that they have it in their power to insist on seeing the sweet sun . . .
Let the people be taught that sunshine means health, joy, the sight of their eyes, and abundance of days; that it is their wealth . . .[34]

THE STEAM DRAGON

On a glorious summer's day in late July 1901, a year and a half after Ruskin's death, Rawnsley rode the "steam dragon" to Coniston. He was coming to visit the exhibition of Ruskin's drawings organized by W. G. Collingwood at the Coniston Institute, and to pay homage at his master's grave in St. Andrews's churchyard. Long before the furor of the Ambleside railway scheme, it had in fact been possible to travel by rail to Coniston. The line was opened in 1859, more than a decade before Ruskin moved to Coniston. The railroad rose steeply from the coast to traverse Torver Moor, terminating in a station built in Swiss chalet style on the south side of Church Beck. The line was constructed in part to ship copper from the mines, but also to increase the flow of tourists in the area. The gondola on Coniston Water was introduced at the same time as the track. During his years in Coniston, Ruskin seems to have used the train many times, though the few references to it in *Fors Clavigera* were characteristically dour and disapproving. On an excursion to see the ruins of Furness Abbey with two protégées, Ruskin complained that the nearby train station was a great eyesore, so ingeniously placed that tourists could see the building "and nothing else but it, through the east window of the Abbot's Chapel, over the ruined altar." Ruskin and his two female wards also had to share the train on the return journey with a group of drunken workmen who had spent the Sunday in the tavern next to the abbey: "It was no use trying to make such men admire the Abbey . . . they were quite an unmanageable sort of people, and had been so for generations." In another letter, he observed that the farmers of Coniston now traveled by train to Ulverston on market days. In years past, the farmers walked twelve miles along the lake and sea, finding refreshment in the clear water from the streams, and enjoying the sights and sounds of the pastoral landscape on the way. But in the present, the traveler paid dear money to travel in comfort, bantering and drinking. He arrived in Ulverston "idle, dusty [and] stupid."[35]

On Rawnsley's train journey to Coniston, there were no such distractions or regrets. When he described the trip to his readers in the 1902 book

Ruskin and the English Lakes, his account waxed both lyrical and learned. The air was "filled with the scent of . . . hay swathes and fragrant from the patches of meadow-sweet . . . left in the field." Passing the coastal towns along the Solway Firth on the way to Foxfield Junction, Rawnsley imagined the hidden histories of the landscape. In the old harbor of Ravenglass, he "dreamed of the days when the Roman general . . . looked out" across the waters at the "western limits of his world." On the Coniston line, he admired the emerald hay meadows, "oaken copses, and pleasant cottages, with the nasturtium already bright in color." The small stations along the line were so "unpretentious" that they almost melted into the "fellside woodland." Deeper and deeper into Ruskin's country they traveled, until they could see Brantwood across Coniston Water. The travelers' eyes caught the "rosy light upon the heather of the higher fells." At the end of the line in Coniston, they left the train "suddenly transported into the world of mountain quietude." They "turned out upon the fellside breast" with a "farm-cottage close by," a "magnificent Scotch fir [standing] darkly against the sunshine on Long Crags," and a "streamlet [murmuring] in our ears." The train and the stations were all essential parts of the sensuous journey into the heart of Ruskin's world. For Rawnsley, steam technology was no longer opposed to the preservation of the Lake District, but a pastoral experience in itself. This was perhaps Rawnsley's main contribution to Ruskin's preservationism. By celebrating the fit between new technology and the environment, he had discovered a way beyond Ruskin's apocalypse.[36]

CHAPTER FIVE

Insatiable Imagination

In the middle of the lake was a small island. From the wooded shores, it looked like a ship at anchor, with its steep sides and a rocky promontory for a stern. The two refugees spotted it just in time. Summer was over. They needed to find a sanctuary before winter began in earnest. The forest to the south swarmed with hostile men. Thorstein was an outlaw from his own kin, and his bride had been cast out from her family too. There was no possibility of hiding among the farmsteads of the Viking settlers or the rude hovels of the aboriginal fell folk. They might have found shelter in Dublin or York, but Thorstein found little to his liking in those crowded and filthy towns. By the time night fell on the lake, he and Raineach had reached the shore and could swim the short distance across the bay together. Though they had managed to carry with them very few things, the island offered an adequate shelter of branches and a small fire. That first evening, they lay together gazing at the moon, feeling like children in their happiness. A howling wind swept down from the fells through the treetops, but the great noise, strange to say, "only made their peace more peaceful."[1]

This fantasy of escape appears in *Thorstein of the Mere*, William Gershom Collingwood's forgotten historical novel from 1895. The story blended archaeology, etymology, and saga literature to imagine the world of Viking settlement in Cumbria around the year 1000. Woven into this swashbuckling adventure were key ideas of the Lake District Arts and Crafts movement: careful descriptions of skilled labor, lovingly rendered landscapes, and a stirring defense of the virtues of the simple life. Elaborate woodcut illustrations accompanied the narrative. Like the utopian fiction of William

Morris, Collingwood's historical romance sought to transmute Ruskin's critique of modern society into an appealing popular form. There was also an element of veiled autobiography in the novel. Like Thorstein, Collingwood had fled to the country with his capable bride, Edith Mary Isaac (Dorrie), by his side, both of them refugees from the horrors of bourgeois suburbia. Just before he was married, Collingwood had experienced an unsettling breakdown in a rented room in Liverpool. He awakened early one morning in the throes of a mysterious illness that precipitated a frightening collapse. In later years, Collingwood would complain of fierce headaches caused by foul air in cities. He found urban life oppressive. Even the seaside resorts were "crowded up and clouded over with the smoke of all Lancashire and Yorkshire." This sense of pollution was moral as much as physical. Crowds milled about the cities aimlessly, wasting time, entranced by tawdry desires and useless objects.[2]

Collingwood's distaste for urban life had been sharpened by his university education. William Gershom Collingwood was born in 1854 to William Collingwood, a painter living in Liverpool, and his wife, Marie. He went up to University College, Oxford, in 1872, where he met John Ruskin and became a favored student: "one of the best and dearest of those Oxford pupils." By all accounts, Collingwood was both charismatic and gifted. He was broad-shouldered and strong, yet fine-boned. His hair was "fair and curly," and like Ruskin he possessed "piercing blue" eyes. After obtaining a First in *Literae Humaniores*, he studied at the Slade School of Fine Arts in London and exhibited at the Royal Academy in 1880. Eventually, Collingwood followed Ruskin to the Lake District to work as his personal secretary and jack-of-all-trades. The area suited his penchant for outdoor activity, including climbing, swimming, and especially walking. Collingwood's 1883 book *The Philosophy of Ornament*—an introduction to the history of the decorative arts—closely echoed his master's views. Like Ruskin, he refused to believe that good decorative arts could be produced by workers enslaved by a mechanized, profit-obsessed society.[3]

Collingwood saw no reason to risk his own or his family's happiness by exposing them to the moral and material corruption of city life. But the move to the Lakes was fraught with practical difficulties. The story of the Collingwoods is a tale neither of tragedy, nor of easy contentment. It took hard work and an appetite for risk to practice the ideal of the simple life suggested by Ruskin. Although Collingwood's family life may at times seem serene and bucolic, financial strain dogged them almost constantly. He was forced to walk the fine line between creative freedom and hardship in order to provide basic comforts for a wife and four children.

Collingwood was aware all the while that something greater was at stake than his personal happiness. Much of his life was dedicated to translating Ruskin's ideas into a different form for a new era. He took vital elements of Ruskin's political economy and refashioned them, marketed them even, to a widening circle of friends, neighbors, tourists, and readers. Rapid economic changes made the task increasingly urgent. The refuge in the north to which he had escaped was already under threat from new industries, hotels, and quarries. From his front window at Lanehead, he could watch detritus spilling down the face of the Old Man. Collingwood fought against such commercial and environmental pressures on several fronts. He furnished his home with local handmade furniture from his friend the Quaker artisan Arthur Simpson. Like Hardwicke Rawnsley, he supported spinning and weaving in the area and even bought a loom with the hope of making tapestries. He engaged in archaeological and antiquarian research in order to illuminate the Lake District's past and to strengthen a sense of local responsibility toward the preservation of the landscape. One of the places he excavated was Peel Island, the tiny ship-shaped rock in Coniston Water that became the home of Thorstein and Raineach in Collingwood's first novel.[4]

It was Collingwood's personal commitment to fostering the good life for his family that would be his major legacy, but only in a way that Collingwood could never have anticipated. His friend and "foster son" Arthur Ransome also wrote a novel, in fact one that was far more famous. Enchanted by the life of the Collingwood family, Ransome invoked the charms of the Lanehead household in his much-loved children's books, beginning with *Swallows and Amazons*. By turning the experience of the Collingwood family into literature, Ransome unwittingly celebrated the legacy of Ruskin in the Lake District. His accounts of rural life and skilled play gave new life to the ethics of consumption among generations of children and adults who had never heard of Ruskin.[5]

AN ESCAPE TO CUMBRIA

Collingwood's quest for the good life began and ended at home. In early 1883 he and Dorrie knew they wanted to marry but did not have the money to do so. Collingwood complained to Dorrie that he was "no good" and that their love was "worth nothing at all," since he lacked the means to support her. Dorrie, dark haired with broad, soft features and a talented painter herself, maintained an almost unwarranted optimism. When she mentioned a house in Orpington that they might take one day, Collingwood was forced to spell out his situation even more clearly, saying it was not a matter of "living cheaply, but of living at all!" It was through Ruskin that Collingwood had the good fortune to meet Susanna Beever, the spinster who lived across Coniston Water at the Thwaite. A generous financial gift from Beever made Collingwood's marriage to Dorrie possible at the end of 1883. A more conventional couple might have settled for life in the city. Collingwood had a decent chance of finding a teaching post and some prospect of a stable livelihood. But that was not the life that he and Dorrie chose. Collingwood veered from the secure and respectable path made possible by his Oxford degree to set up shop instead as a landscape painter in Cumbria. Dorrie's feelings were tested as they approached their wedding date, but she never seems to have wavered. In William, she found a life beyond the ordinary, an experience richer and freer than what a less adventurous man could have offered her. The life they ultimately made together could, in some ways, be called "simple." Yet it abounded in artistic and intellectual richness, built on a radical environmental and moral premise.[6]

At the heart of this experiment was the faculty of the imagination. Modern economic theory insists that human desires are insatiable. People always want more than they can have. This, economists assert, is a cardinal feature of human nature, and a wonderful gift to mankind. Under the right social and political conditions, this kind of insatiability will be the driver of long-term economic growth. Critics, however, argue that consumer society is the product of a historically specific social system that is ecologically unsustainable in the long run. We might also think of human desire as a force that can take a wide variety of objects: public, private, economic, social, emotional, and artistic. The creative imagination expands the possibilities of fulfillment toward a range of objects that have a relatively low economic cost and small environmental impact. At the same time, it opens the door to a social life beyond the market, oriented toward people rather than consumer goods. On this count, the imagination becomes a counterforce against ordinary consumption. A materially simple life can still be quite rich. This, at any rate, seems to have been the wager of the Collingwood family.[7]

Yet their experiment easily could have stumbled at the outset. That the Collingwood family nearly failed to materialize is an understatement. Collingwood felt great anxiety in taking Dorrie away from her parents and her established social circle in Maldon, Essex. He needed to make sure she understood what sort of life he had in mind. When he received the happy news of Beever's gift in 1883, Collingwood wrote Dorrie a long, detailed letter regarding the difficult road ahead and the plan to move into the ramshackle cottage known as Gillhead on the shore of Windermere in the Lake

District (Collingwood had spent his school holidays there). He was overjoyed when she accepted the proposal, and playfully noted, "I hope I'll not disappoint you <u>much</u>, when I get you into the bear's den." But after that, the letter turned rather practical. She would have to learn the area's geography; he would go to the registrar (a "long walk" of ten or twelve miles—he had no horse); and she would have to get new clothes—though, he added, "Don't worry about a great wardrobe." She would need only a "macintosh, & unspoilable hat" and india rubber boots, preferably much too large so she could fit extra socks in them in winter. It was more important that she should be able to "take nice walks" than to strike an elegant figure. One cannot help but wonder at the young woman consenting to leave Maldon and its proximity to London for a rundown Windermere cottage—a home that had once belonged to Collingwood's childhood nurse. He told her there were some pleasant acquaintances to visit in Bowness, the closest town. But there were no carriages or cabs, and thus evening visits would be out of the question. Perhaps Dorrie liked the idea of being whisked away to the Lakes, but Collingwood himself often seemed far from romantic. His obvious passion and excitement were always tempered by hard calculations. Much of the engagement letter dealt directly with financial matters—what they had in savings, what everything would cost, what they could expect by way of wedding gifts . . . even down to the forks and knives. There was little hope of extravagant living:[8]

> Now I must tell you about money—I have £5 in hand: £40 in bank: £6 on some of my copies of [*illegible*] paid for: & you know what in the bush: and the bush, it seems, is to be beaten about Xmas.
> My living here will use up the £5
> My [house—repair]   £10
> Travelling & my clothes   £10
> Leaving £26 for furniture—carriage of it—& living, until more comes in. That is very little isn't it.

The wedding itself gave him a particular headache, enough to reject any notion of expensive festivities. "There will of course be no entertainment," he told Dorrie. If his brother and sister wished to see a finer affair, he wrote, they would have to "do it on their own book." He reluctantly agreed to purchase a "frock coat and top hat"—but only because it might be useful for future "occasions of solemnity." He even warned her not to expect "anything grand" in the way of a wedding gift from him. The niceties of civilized society did not fit his budget or his moral or social code—and they never really had: "I'm as ignorant as a clod about all such things." Soon he lost patience with the whole affair, small as it was, gasping, "Seriously—are we expected to have a cake?—what childishness!"[9]

Did Dorrie really understand what she was going to give up? Or what she was heading toward? Perhaps not entirely. Collingwood continued to spell out for her the realities of the life they faced, as though he couldn't be sure she had grasped what a change of status she was in for. He held up no illusions. They would live (for seven years, as it turned out) in a "poor old house . . . a sad rotten concern" that would have to do "till we can afford a better or have to retire into the workhouse." At Gillhead, he told her, she would have no one to turn to in a crisis but neighboring cottagers. Indeed, she would have to learn the local dialect to get by. He wondered several times whether she would despair at "the poorness of the place . . . and the loneliness." He even confessed he wished sometimes she had "taken a safer man."[10]

Yet Dorrie seems to have taken each warning in stride. She loved Collingwood, art, and the countryside, and her choice was made. She knew she could hardly separate him from the landscape he loved. Collingwood and his father had visited and painted the Lakes for more than a decade by 1883, and as hesitant as he sometimes was, in general he could barely restrain his enthusiasm: "Won't you like the place! Yesterday afternoon I paddled up to Bowness & round the big Island. It was fine & calm: and the gloriousness of everything was quite bewildering." He assured her it was not merely picturesque, but exciting as well, for they could reach the bigger hills at any time. If he could not offer her a comfortable house and security, he offered her true adventure and partnership. Collingwood was quite egalitarian for his time, even urging Dorrie to keep her own bank account: "I want you to be independent in such things; it saves trouble in all." He sought to foster freedom, of a late Victorian sort, within his own family and in the community around him. He wanted each person to have the space and means to flourish, and the autonomy to act as they deemed best. This freedom included imagination, industry, confidence, and perhaps most of all self-reliance. If he was not the "safest" man for her to marry, he meant to teach her useful skills so she could keep herself and her loved ones happy and secure . . . or at least assist him in crawling back from the edge of this or that financial abyss. Such self-reliance, Collingwood believed, grew out of the very landscape, the rugged fells of the Lake District, and out of the people who had lived there for centuries, going back to the Vikings and beyond. This lovely, difficult land cleared away distractions, focusing the mind on what was truly important.[11]

Collingwood's decision to take up residence in the Lake District was not an easy choice. Obviously he loved the countryside and found he could just manage a life there as Ruskin's secretary, a painter, and a writer. Yet he had deliberately eschewed a very plausible alternative life as a scholar. Collingwood studied classics and philosophy at Oxford in the early 1870s, and despite his mother's death from consumption in his first year, he did exceptionally well. He won a First on his exams in 1876, along with the Lothian Prize for an essay he wrote. The master of his college thought very highly of him, calling him "'one of the ablest and most thoughtful'" students he had ever met. With such accolades, his future looked very bright. He might have spent his entire life in the comfortable circumstances of an Oxford don. Collingwood's father even pressured him to apply for scholarships, though he had already taken up painting at the Slade School of Fine Art. Unfortunately the double effort seems to have caused him to fail on both accounts. He did not win any scholarships, and complained, "I spoilt all my [art] work there by trying for fellowships, as my father wanted, & dividing my attention."[12]

Collingwood's career as a painter was uneven, and his confidence was often tested. After attending the Slade, he tried to get by in London as a painter and art tutor, but found that he could not earn enough and so lived a life he deemed a "purgatory." He wondered about his talent too. In 1889 he wrote to Dorrie of his fear that "I can't paint after all. But I go up & down about that, as you know." Although he did carry on with painting throughout his life, he took on a huge variety of other pursuits as well, including archaeology, poetry, historical fiction, decorative arts, Ruskin's biography, and a tourist guide to the Lake District and the Coniston area. In the end, Collingwood's talent as a painter may not have been as striking as he had once hoped. Yet he traded the likely security and status of a solid academic career for this piecemeal life full of financial pressures, social isolation, and artistic doubt. Why? A good guess is that Collingwood used his ability to paint conventional portraits and landscapes quite deliberately to help facilitate the "good life" that was his true goal. The whole was greater than the sum of its parts.[13]

When Collingwood met Ruskin as a young student at Oxford, he became involved in the famous Hinksey Dig. There, young well-to-do scholars were directed by Ruskin to build a road to a nearby village as an act of charity and muscular virtue, although Ruskin's future private secretary was a little "slack" at the dig, he admitted. But the lesson of Hinksey was foisted on him once again in 1875, when he and another student, Alexander Wedderburn, came to Brantwood to work on Xenophon's *Economist*. Ruskin thought the ancient Greek treatise laid out the best means of managing one's household and suggested a philosophical definition of wealth in terms of order, self-restraint, and the satisfaction of simple needs. The translation was dedicated to all "British peasants." Ruskin hoped to print it cheaply enough that it could be owned by every farmhand and shepherd. The two young men spent half the day translating and the other half shoveling dirt and lifting stones to build a small dock on the beach below Brantwood. Though Ruskin's hopes to diffuse Greek virtue among the rural poor proved elusive, Collingwood at least took the lesson to heart.[14]

This idealism is evident in Collingwood's first publication of his own, *The Philosophy of Ornament* (1883), which gathered together lectures he had given at University College in Liverpool on the history of decorative art. The chapter on prehistory described with great admiration the thighbone of a reindeer whittled by a cave dweller ages ago. On the back end of the short, sharp stake, there was also a reindeer's head—a purely decorative carving. Utility and beauty had been crafted together in a seamless, organic manner, even at the dawn of human history. Modern societies were far less likely to produce such marvelous objects: "Indeed, civilization rather hinders than helps the craftsman." Near the end of the book, Collingwood spelled out how capitalist large-scale production endangered artistic freedom: "[A] real national style can only exist when the people produce it, and . . . enjoy it . . . when the people can have artistic, as opposed to mechanical, lives." The decorative arts, then, could not be great if the nation endorsed the cheap mass production of goods. Industrialization, Collingwood thought, "not merely permits but promotes organized resistance to individuality on one hand, and on the other to noble exertion or emotion." Creativity, freedom, and spontaneity were tied to the individual's hands. But the laborer's life had also to include rest and time to make beautiful objects and enjoy their work.[15]

CONSUMING WISELY

Married in late 1883 and living at Gillhead, the Collingwoods settled into their rural life in Windermere. They had their first child, Dora, in 1886, and a year later, Barbara. At this time Collingwood had begun taking part in Arts and Crafts exhibitions put on by Arthur Simpson, a wood-carver and furniture maker in Kendal. Simpson was intensely interested in Ruskin's critique of modern industry. He especially appreciated the idea of dignity in work and the allied notion that art should draw its power and delight from imitating the divine design of the natural world. Since Ruskin was already gravely ill and often unresponsive when Simpson came to visit Brantwood, Collingwood became his principal teacher and interlocutor. Like Rawnsley and Fleming, the furniture maker saw manual work as a way to virtue rather than profit. He enjoined his wood-carving students to give up the idea of making money with their craft, and to focus on the "pleasure of it" instead. "Without this love," he told them, "your efforts are likely to be futile." He and Collingwood were in agreement on this point. Nevertheless, Collingwood spoke frankly to him about making a living with his craft in the Lakes region. When Simpson's exhibition generated sales of his and Dorrie's paintings, Collingwood was thrilled. "You are really doing capitally for us." Then he admitted, "Much wants more — and now I am greedy enough to wish to be able to say that we have sold £50 worth of pictures! . . . But I am very content with what we've got." Such contentment was difficult to maintain.[16]

The Collingwoods teetered on the edge of bankruptcy throughout their time at Gillhead. But Susanna Beever, who had already given them money to marry in 1883, came to the rescue once more in 1887 with a £100 birthday present, averting another crisis: "Coming on the top of her other present and several little sums I have made & which already helped us to turn the corner which Ruthie [Collingwood's sister] was so much dreading — now puts us into affluence." Somewhat defensively, Collingwood stressed that he had "already" made good himself, and that Beever's gift simply helped them live well. Yet he lingered over her generosity, stressing the fact that while she *could* have made him feel beholden, she had not. This must have been an uncomfortable thought, especially coupled with Ruthie's worries. At any rate, a "corner" had been turned, once more.[17]

In 1891 the Collingwoods — including the youngest children, Robin (b. 1889) and Ursula (b. 1891) — moved to a larger house called Lanehead, situated just a mile north of Ruskin's lakeside home, Brantwood. It was far more convenient for Collingwood's work with Ruskin, since he could now avoid the long trip from Windermere. Yet he was embarrassed at Lanehead's size. He told his father that he "would have taken a smaller one if I could have found it," but too many factors led him to accept the rental agreement on the house. Strings were clearly pulled to enable the family to live there. As one scholar points out, at the beginning of the 1890s Ruskin paid Collingwood exactly what he owed in rent for Lanehead; this in turn amounted to one-third less than earlier occupants had paid to live in the

FIGURE 5.1 The Collingwood family home, Lanehead, with WGC and one of his daughters (either Dora or Barbara) seated, ca. 1903. The family usually exited to the lakeside lawn through the window (note the steps below it). Teresa Smith.

house. Although Collingwood appreciated his new home, it still felt like an extravagance. He told his father wryly, "It will amuse you to see us being grand people at last: for our house is one of the biggest—with stables and conservatories and all the rest of the parade of gentility." As it happens, those great stables never housed more than a single donkey during all their time in the house.[18]

Though Collingwood had not intended to live in such a house, the decision to furnish it in style was entirely his own. He lost no time writing to Arthur Simpson, and began the process of fitting out his new home and designing his own furniture. In letter after letter over several years, he discussed the details of these plans with his friend. Sometimes he worried that his directions were too amateurish to make sense. The total cost too was uncertain. He asked Simpson to send his wife a "surprise" of a "Scandinavian Sofa"—but only if it would not be "extravagantly expensive." Collingwood seems to have been caught between the prudence of the good household "economist" and his own fascination with design. Purchasing quality goods from a skilled artisan mattered to him, so he was careful to negotiate the terms of work to reach the best possible outcome. But the process proved full of surprises, not all of them pleasant. Regarding the floor of his house, Collingwood complained, "I am sorry you or Tom did not [candidly] tell me that the floor wouldn't make a good job stained. Tom's colour . . . has turned out less of a green than a dirty brown; and his stippling being clumsily lumped on . . . looks like the ink spots of a schoolroom floor." But in this case no amount of grumbling could undo the damage; Collingwood had to resign himself to covering it with carpet. In the same letter, he griped about the process of designing furniture from scratch without any comparable projects to study: "It is a little handicapping to be asked to decide on so important & conspicuous a piece of furniture from only one example." He vowed to look up grates similar to the one he wanted on his next trip to London. In his frustration, Collingwood of all people began to hanker after ready-made goods on offer in the metropolis. Playing the prudent consumer in a rural backwater, with limited or no inventories, proved both annoying and challenging.[19]

Yet Collingwood managed to maintain a warm relationship with his artisan friend. Simpson sent a further set of designs later in 1891. Collingwood responded with six detailed sketches of his own, containing numerous descriptions and suggested modifications. At issue were the "settle," the "writing table—man's cabinet & corner cupboard," and a small table. Only Simpson's large dining table design was accepted without alteration. Although Simpson was capable of adorning works with intricately carved floral motifs in the arts and crafts style, Collingwood requested a rather plain design marked by strong vertical lines. He said "the general proportions" were "more important than the detail," yet he specified that each of the pieces were to have a miniature decorative balustrade along the top or back of it. Further, the tables were to be "as black as possible," while Dorrie requested that the "green things" should be "made of ash & not of oak." The aim was a set of furniture with simple lines that commanded attention because of its size and aesthetic coherence—not to mention the fashionable green coloring. Collingwood's fastidious amateur descriptions may have created unexpected problems for Simpson. But given time, patience, and a commitment to working things out (and an ability to accept defeat on occasion), they came to an understanding. There is no record of the total expense, but it may have cost more than ready-made furniture from the cities. It certainly took more time and effort to design.[20]

One might wonder why Collingwood did not make do with older pieces of furniture, rather than commissioning or purchasing all new ones. Ruskin himself brought much of his parents' furniture to Brantwood. Ruskin had complained in *The Poetry of Architecture* (1837–38) that England's "perpetually increasing prosperity and active enterprise" killed any appreciation for well-made, durable objects: "nothing is allowed to remain till it gets old. Large old trees are cut down for timber; old houses are pulled down for the materials; and old furniture is laughed at and neglected." It is possible that Collingwood's father, also a painter, had little or no furniture to spare, and in any case still had need of it himself. As an antiquarian and archaeologist, Collingwood may have relished the idea of buying antiques, a practice that was becoming more common by the time he moved to the Lake District in 1883. But, like Ruskin, he also supported the notion of craftsmen finding enough work to earn their living from a job well done.[21]

Despite Collingwood's great effort at fitting out Lanehead, the house never entirely suited him. It was apparently too conservative, too grand for his taste. "When we get our ideal house," he told Dorrie, "let us not have a lawn in front of the study windows, but some woodland & distant view." He confessed that he still looked back "to Gillhead with hankering"—the dilapidated cottage where the couple had spent the early years of their marriage. This yearning for simplicity was entirely in keeping with Collingwood's many projects in Coniston.[22]

THE PRACTICAL UTOPIAN

At times Collingwood was forced to exercise his more practical side, even with regard to furnishing. His salary as a secretary to Ruskin was apparently spent entirely on rent. Everything else he needed to support a family of six depended on his and Dorrie's paintings and Collingwood's writing. In a series of letters to an unnamed correspondent who wished to develop an arts club in the Lake District, Collingwood provided worldly advice. He was completely prosaic about his vocation, commenting, "there is a market for good cheap work all over the district—and if the artists don't combine to work the market, it is their own loss." He observed that his own typical sales were not to people in "the wealthy class," but rather his neighbors. Bluntly, he explained that he did not "like to talk cant about doing good by spreading art among the people." He painted to make a living; if his less educated neighbors benefited from it, that was good, but not the primary motive. A similar concern for the business end of artistic life showed through in his letters to Arthur Simpson. Simpson lacked Collingwood's social standing, polish, and education, and having moved to Kendal recently to start his business, he sought Collingwood's advice on a train-

FIGURE 5.2 William Gershom Collingwood. Teresa Smith.

ing program he was to oversee. Collingwood reviewed the prospectus and commented succinctly, "Quite right not to mention terms—you may vary them according to requirement of tuition & accommodations." He pointed out that Simpson's wife would be an added attraction to "both men & girl students,—the men wd know that they could get their dinners, & the girls that their morals would be looked after, if a married lady is on the spot." Collingwood always encouraged women to join local art and archaeological clubs; his own wife and daughters were artists. But he also saw women, simply put, as good for business.[23]

On the one hand, then, Collingwood's habits and lifestyle suggested an idealistic, anti-industrial frame of mind. But on the other hand, his life as a

Lakeland country writer depended on hardheaded calculations. Later on, when his own young son showed signs of gentleness and sensitivity, he remarked to his wife, "It is nice that Rob has humanity in him: I hope he won't be too spoony." To live the good life, he knew, one had to be somewhat pragmatic. William and Dorrie kept a fairly close eye on household expenses. As one scholar notes, he "did not possess private means."[24] For years their income depended on how many paintings or books they could sell, as well as on Collingwood's lectures and teaching. Much of the food consumed at Lanehead seems to have been local or regional in origin, like the table kept at Brantwood. The pinched state of the family finances probably also contributed to this diet. Ursula's drawing of a jam jar with the label "Best Home Made" tells us the children were adopting a critical view of consumer desires even at a young age—perhaps also learning to like what was grown nearby, and free.[25] Sometimes Collingwood worried openly that despite his shrewd approach to art, he might not make enough to pull through the year. He confessed to Dorrie at one point that he felt foolish at times to work so hard on speculation. Writing books and painting pictures required an enormous outlay of time and energy, for no certain return. The continual sense of risk unnerved him. Even his children caught some of the family fear, along with a stubborn adherence to their own dreams. Dora, the eldest daughter, began painting with a sure sense of all its excitement and sacrifice. An undated record of her daily purchases reveals her frugality, with every last penny accounted for. She included everything from butter, coffee, eggs, and apples to "hooks & eyes" (for sewing), canvas, and matches. In 1907 his youngest daughter, Ursula, wrote to her mother: "I am so glad Daddy has sold that picture and that you are all right." It seems the fate of the anti-industrial painter's family depended on the whims of the market and the idiosyncratic taste of buyers.[26]

To make matters worse, Dorrie suffered a serious foot ailment in 1893, and was forced to go south for her convalescence. Collingwood was left in charge of the household, including a cook, a housemaid, and three of the four children. His letters gave Dorrie frequent updates on the weather, errands, social visits, money, and his work. Sometimes he shared a few, very gentle complaints, as when he was up all night because Barbara was sick or when unexpected guests stayed two hours (though he had no idea who they were). He also dealt with other basic chores, such as getting the family's donkey shod or the mail cart fixed. His sister, Ruthie, stayed with him for a time, but seems to have been more of a bother than a help. It was a frustrating month. Even the donkey led to delicate financial worries. He asked

Dorrie to inquire of her father "whether one must pay tax on a donkey carriage—I hardly like asking at Brantwood (!)." It could not have been easy to get all of his work done under such circumstances, and it was certainly not the only time Dorrie left him alone to manage everything.[27]

Robin Collingwood's education at Rugby and Oxford was paid for not by his father, but by Emma Holt, a family friend. If Collingwood lacked private wealth, he often appears to have made up for it with social connections, not to mention the force of his principles and convictions. As for his daughters, in 1905 Collingwood took on "the first, and only, full-time employment of his career"—an appointment at University College, Reading, where he lectured on fine art. He grumbled about the move away from Coniston and resigned with great relief in 1911. But the job helped pay for the education of Dora and Barbara. Collingwood's salary was £300–350 per year. This might have provided a means of accruing significant savings, if he had stayed on. But instead he went back to his preferred way of life, which included a great deal of unpaid work for the Cumberland and Westmoreland Antiquarian and Archaeological Society as editor and later as president. Collingwood wryly wondered if anyone had ever been "so busy, for such a trifling return in cash." At the same time, club life was in itself an expensive pursuit. The annual subscriptions he paid to a wide variety of societies and clubs—many of them antiquarian or artistic—amounted to a fair sum, between four and seven guineas per year.[28]

Nearly all of Collingwood's projects sought to promote the legacy of Ruskin in the Lakes region and beyond. But for the most part, he avoided overt political activism. Other defenders of the Lakes, like Hardwicke Rawnsley and Gordon Wordsworth (William Wordsworth's grandson), were far more suited to that task.[29] Instead, Collingwood preferred to act through the medium of his writings and public talks. "I don't think I shall be in my right place as chairman on a committee or anything of that sort," he told his friend Arthur Marwick. "My business is to write lectures, not to organize." In addition to the numerous unpaid articles on archaeology, he tried his hand at biography, historical fiction, and local guidebooks. The most lucrative venture of this sort was the biography of Ruskin, first published in 1893, which brought in a solid income to the family for many years, much as *Frondes Agrestes* had bolstered Susanna Beever's finances. However, Collingwood declined to collaborate on the Ruskin Library Edition with Alexander Wedderburn and Edward Tyas Cook. Rather bizarrely, he felt that this eminent work of scholarship was merely a mercenary business venture by the executors. But the rejection no doubt preserved time for other commitments.[30]

He spent about two months every year in London painting portraits, finding this just remunerative enough to last them through the rest of the year in Coniston. It is difficult to say whether Collingwood's painting is better or worse than his reputation as a modestly talented portraitist and purveyor of conventional views of the Lake District. Ruskin characterized his work as "accomplished and amiable"—a patronizing and anemic sort of praise. Perhaps the master felt obligated to commend his devoted personal secretary. Collingwood's own remarks about making money as an artist and his strategy of selling cheap pieces to less educated neighbors indicates a certain awareness of the limitations of his talents. Surprisingly, his family letters from this period indicate little disappointment about the matter; indeed, he made very few jaded, weary, or ironic comments on artistic matters, except for the occasional reference to some of his works as "potboilers." Obviously Collingwood found enough distractions and pleasures in rural life to make light of any disappointments about professional security or artistic fame. In a letter to his father, he expressed some regret that neither he nor his brother, David, had much luck in business. "We don't seem to be a money keeping lot," he lamented. Dorrie had been urging him to "begin to save," but he felt all he could ever manage was breaking even. Characteristically, Collingwood ended his financial musings with a quick note about the glorious weather. They were all "sunburnt and in flourishing health."[31]

VIKINGS AT THIRLMERE

Yet life in Arcadia was not secure. Like his friends Susanna Beever and Hardwicke Rawnsley, Collingwood understood all too well that the natural beauty of the Lake District was fragile and under siege. In 1894 the region was already witnessing environmental change on a grand scale. Manchester Corporation had fought and won the two-decade battle to dam and pipe the waters of Thirlmere into Manchester. Thirlmere was at that time covered with lush deciduous woodlands. But it was thought that deciduous trees would cause problems when they shed their leaves into the water system. Therefore, the forests around Thirlmere's sparkling waters began to be replaced with a dark conifer wood consisting of pines, spruces, fir, and larch, uncharacteristic of the Lakes region. Enormous boulders remained, lying scattered about the shore. Harsh, rocky outcrops rose above the tree line. The sheer face of Raven Crag stood guard at the water's edge, and the ridge of Helvellyn, one of the three tallest mountains in the Lake District, stretched along the other side of the lake. The area captured Collingwood's imagination. He speculated that the Viking settlers must have held their annual Althing (assembly) at Legburthwaite just north of the lake.[32]

Thirlmere was not originally one lake; it was two bodies of water connected by a narrow strip. An ancient and picturesque "Celtic" bridge consisting of large piles of stones joined by a few wooden planks connected the two shores at the narrowest point. Once dammed, the water level rose dramatically and the ancient bridge disappeared beneath it as the two lakes became one. The lakeside settlements of Armboth and Wythburn disappeared, along with their roads and cottages. Fish stocks, trees, and all manner of wildlife were also irrevocably changed. In just a few years, the original mysterious and sublime landscape became nothing but a memory. In 1902 Collingwood wrote in his tourist guide *The Lake Counties*, "Thirlmere has no expanse, but it once was the richest in story and scenery of all the Lakes. The old charm of its shores has quite vanished, and the sites of its legends are hopelessly altered."[33]

In the same year the dam was completed, Collingwood was finishing the first draft of his historical novel *Thorstein of the Mere* (1895). Like the fiction of J. R. R. Tolkien a generation later, Collingwood's world arose from a deep delight with etymology and language. Why, he asked, was the first recorded name for Coniston Water the mere of Thurston? Who was this Thorstein? The question of the name then set Collingwood's imagination to work. What was it like to live in Cumbria when it was still a forested wilderness? Out of this conceit, Collingwood spun a delicious yarn. Young Thorstein was kidnapped by giants, but managed to escape, was subjected to treacherous machinations on the part of his brothers, married the wrong girl (who was refined but weak spirited), went in search of his true beloved (an uncouth giant girl), escaped again with her to a secluded island, and finally died in battle while defending his people and their way of life. Collingwood's story was "multilingual and multicultural," depicting the societies of Viking settlers, Celtic hill folk, and Anglo-Saxon Lowlanders. Thirlmere appeared three times in the narrative. At the Thing Mount nearby, the Viking settlers convened to settle disputes and uphold their liberty. On its shores stood the castle rock where Thorstein was kept in captivity. Thirlmere was also the place where the hero was killed in the final great battle between the Saxons and Vikings. The plot offered a frolicking romp through England's Viking past that rivaled the thrill and wonder of Robert Louis Stevenson's *Kidnapped* (1886). Not surprisingly, Collingwood's son, Robin, adored the book. The detailed descriptions of the landscape offered a new angle on the

FIGURE 5.3 Title page of *Thorstein of the Mere*, by W. G. Collingwood, 1895.

sparser narrative structure of the Icelandic sagas, upon which the novel was based. Adventures were set in a time "when all the mills and houses were unbuilt, and the land uncleared, and nothing but wild timber, dank and dense, filled the dale, with the logs that rotted where they fell . . . and wild bulls and wild boars, wolves and cats, hag-worms and lizards, and maybe a bear or two." There, the Northmen made their homes on a frontier of deep woodlands.[34]

But the novel also offered signposts for the future. Collingwood emphasized that what had been lost was, in a sense, a Ruskinian way of life—and one that could be resurrected. The novel abounded with an appreciation for manual skills, and lingered on technical descriptions of survival in a harsh environment. Some of the objects of Viking society were pictured in Collingwood's arts and crafts images specially designed for the book. Young Thorstein and his brothers grew up learning to take care of themselves and to make useful, beautiful household goods: "They were not brought up in idleness," and despite the presence of household servants and serfs, "it was the way of these people to do their own work." They were proud of their skill with iron, shoeing their animals and making weapons; they learned basic woodworking; and the women mastered spinning and weaving. This was not a savage, hard-scrabble sort of life, but a society that valued the skill and knowledge necessary to live well. There was always time to linger a little over the things they made, adding some "conceit of daintiness," "curling the horns of a door-latch," or "engraving a blade with devices." Throughout, Collingwood drew a sharp distinction between town and country. Urban life in the tenth century was evidently as corrupt and unhealthy as the Victorian cities. When Thorstein visited York, he was shocked at the crowds buying and selling wares, rich and poor alike "all cheek by jowl." The atmosphere was "foul with the refuse and rubbish of a thickly inhabited town." There was chaos and "tumult," and people scurried "like rats out of their holes." Thorstein was taken in by a merchant and his wife. They owned many things "rich and rare"; there were bed "hangings" and "carved work" on tables and stools, "shining copper pots and pans," and the family wore elaborate clothing and jewels. They crammed their goods into a crowded, dingy home with a pretentiousness befitting lords and ladies. Thorstein found nothing to admire in this gauche display. Despite their fine dress, the merchant's children "would have looked but blue and wan alongside the applecheeked rogues from the Northmen's homesteads." One hears over and over in these descriptions Ruskin's mantra: there is no wealth but life.[35]

To this same end, Thorstein's two love interests, the "fresh and fair" Asdis and the large-boned, capable giant Raineach, represented distinct moral perspectives on consumption. Though this was a heavy-handed contrast by the standards of the late nineteenth-century novel, the conceit was fully in keeping with the conventions of the Viking sagas. Asdis sat in her home idly eating "sweet cakes, one after another," falling asleep midday and finding life "dull in these backwoods." Raineach, by contrast, might be "ugly" and filthy, descended from a clan of Celtic savages, but when Thorstein was first abducted by them, he awakened to find her "nursing his head and weeping over him." She was selfless and unassuming, and on multiple occasions

FIGURE 5.4 A map from *Thorstein of the Mere* showing Lakeland as a frontier of Viking settlement. This was the historical origin of the statesman tradition, according to Collingwood.

INSATIABLE IMAGINATION 139

proved her courage in the face of danger. Unlike Asdis, Raineach was utterly satisfied "just to be let alone, and to be together" on the little island to which she escaped with Thorstein. She did not pine away for gossip, goods, or society of any sort. Her contentment with the necessaries of life made her a valuable, self-reliant partner for Thorstein. In a brief interlude, Collingwood described how Thorstein and Raineach built their house using "poles" and "boughs" and "turfs." When they were done, "there was as snug a home as might be in all Lakeland." Their days in Viking Coniston were easy and contented, "as if life were one holiday."[36]

In 1902 Collingwood penned a tourist guide—*The Lake Counties*—that

FIGURE 5.5 Asdis, Thorstein's Viking wife in her Arts and Crafts–style home.

encouraged visitors to explore the hidden histories of Cumbria. The ideal of freedom set out in *Thorstein of the Mere* was here spelled out as historical fact. Since the time of the first Danish and Norse settlements, "Dales-folk" had asserted a certain measure of "independence" from the power of the central government and feudal lords. The common people also enjoyed a great degree of social equality, free from "emulating the grandeurs of town or castle." Better still, by living off the land, their well-being "did not even depend upon the state of the market": "Their life was like that of the Swiss or the Icelanders, where every farm is a separate estate." Every family spun and wove clothes for themselves and grew their own food, "and wanted no more." While "They were forced to be careful," and in a sense lacked consumer choice, this was not taken strictly as a hardship. To Collingwood's mind, the "Dales-folk" still "kept alive some smouldering memory of their birthright." However, the guide left unresolved the question of how present and future development might transform the locals. Could the old way of life be preserved in the modern age?[37]

JOURNEY TO THE PAST

In 1897, just two years after the publication of *Thorstein of the Mere*, Collingwood embarked on a mission of time travel. He went by steamship to Iceland to see what remained of the industrious, creative, free people depicted in the saga literature of the Vikings. The three-month excursion was on the surface a business scheme. He planned to paint all the places mentioned in the sagas, hold an exhibition in London to sell individual pictures, and publish reproductions in book form. He thought public interest might be high, given all the new translations of the sagas into English. But this journey was also a "pilgrimage," an expedition to discover firsthand the parallels he presumed existed between the present-day Icelanders and the Lakeland Northmen of the tenth century. The English Vikings were, after all, descended from the same stock that settled Iceland. The Free State of Iceland had flourished for a few centuries, but liberty had been lost under the yoke of Norwegian kings and Danish colonial rule. Denmark imposed strict regulations on Iceland's economy; thus, the island country was fairly well insulated from the modern world by dint of its poverty. To an anti-industrialist attuned to the joys of simple living, this was not an entirely unfortunate accident of colonialism. Yet the dismay Collingwood conveyed in his letters home indicates that he was quickly forced to discard any illusions about the survival of archaic traits and the ennobling features of Icelandic subsistence.[38]

Collingwood embarked on his journey during the summer months. Before reaching Iceland, the ship made a quick and delightful pass through the "fairy" land of the Faroe Islands. In one of the dozens of letters he wrote home to his wife, children, father, and siblings, he described the greenery around the wooden farmhouses, with "marshmarigolds," turf-roofed houses, "gooseberry and currant bushes," "primroses," and "daisies." The near absence of trees also caught his eye, but the islands had charm enough without them. The houses were set into streets that, he said, "can't be bicycled or driven in," and inside the houses was "such jolly old quaint comfort" with "good furniture and brass things and walls paneled and papered." The Faroe Islands seemed like a reservation for the simple life in the modern world. Everywhere he looked, there were carefully tended gardens, rule-abiding citizens who knew where they could and could not bicycle or drive, and the cultivation of a secluded, peaceful, old-fashioned home life.[39]

Even from a distance, Iceland paled in comparison. Despite the "snow topped mountains," the town of Reykjavik with its "straggling . . . wooden and iron houses" was "poor and mean looking." Due to frequent earthquakes, Icelanders did not build houses with stones, despite having an abundant supply. He griped that "apparently they don't know how to do dry walling, and there are only 2 or 3 masons in the place." Icelanders did not seem to take much care with decorative work: the hotel was "rough" and "painted in ugly colours." The women wore severe caps, and the men had "no particular costume." Collingwood was shocked at how the descendants of the sagas' heroic early settlers had failed to follow their distant ancestors' examples and now wallowed in filth, ignorance, and aesthetic complacency: "Everybody looks poor and the place entirely forlorn and heartbroken with bleakness and neglect and apparent repression (so it seemed to me)." Extra paints had to be shipped to Collingwood, when he discovered that none could be found in Iceland: "The only 'artist' [here] is a poor old man who has made a museum of birds, and eggs, and fishes—and drawn them with penny paints." He was baffled by the dismal standards in local arts and crafts, and could not see why it should be so. In *The Philosophy of Ornament* (1883), Collingwood had admired the beautiful carved horn of a prehistoric artisan—an example of skill, cleverness, and a basic desire for beauty and fitness, obviously created with few, if any, resources. But Iceland lacked all artistic care, as far as he could see. He complained that the "deadly ugly" houses were painted once, but "never again." The Icelanders' apparent apathy and lack of orderly living violated the basic premise of Xenophon's *Economist*. In the ancient Greek text, a well-kept household provided the basis for the economy. True wealth began at home, not in the marketplace—unless one's character prevented it.[40]

Despite this ominous beginning, Collingwood departed on his journey along the coast to the sagasteads. It proved a grueling expedition. He and his Icelandic companion were scorched by the sun during interminable, treeless pony rides. At other times they were treated to freezing temperatures. They suffered through grim meals and cramped boarding facilities. The rooms they secured were often in the private homes of villagers; there, Collingwood fought off lice and suffocated beneath hot, thick duvets that were nevertheless too short to cover his feet. The local guides routinely underestimated distances. Yet, these difficulties aside, or perhaps because of them, his journey proved a great adventure, full of marvels to rival those of Thorstein himself. He wrote home about the sagasteads he was painting, describing the sites of famous happenings—for he was certain that the stories must be based on real events. He told Dora of the killing of Kjartan in the famous *Laxdæla Saga*: "Well, my dear, we found the spot, all just as it had happened—and for the first time these thousand years, drew it just as it was." This meeting with what Collingwood believed was the literal past was enough to make him forget "for a while that one is wet and cold and hungry and miserable." He was happily distracted "in thinking of the people who were there so long ago, and about whom fine books have been made, telling true stories of their doings." Much like Ruskin, Collingwood felt that "the thing that appeals to us all . . . and carries us out of ourselves, is the union of story and scenery." A landscape without the memory and story of humans living within it was meaningless.[41]

The experience in Iceland also seemed to confirm Collingwood's fears about the possibility of permanent environmental deterioration. He encountered there a landscape that appeared to be severely eroded by bad usage over centuries. Here was a vicious circle in which a degraded land debased its inhabitants' character, decreasing any inclination toward wise management of resources. In one ominous passage about an Icelandic farmstead called Saurar, which he privately christened "Sawrey" (the name of a village near Coniston), he linked the fate of the Lake District directly to that of Iceland. He lamented the signs of neglect and environmental destruction: "All this valley and the next valleys and the shore of the fjord used to be full of wood." The landscape reminded him of the barren "top of the Old Man" in Coniston. But Icelanders had even destroyed the lowlands by taking "turf to build their huts and walls," so they became eroded and bare, unable to regenerate native grasses. Sadly, the people of Iceland were even "more

FIGURE 5.6 Ruins from the slate-mining industry on the Old Man of Coniston, 2011.

prejudiced and ignorant" than the local inhabitants of the Lake District. Their profound complacency unnerved him greatly.[42]

Thirty years later, Collingwood was still voicing his concern. In 1924 he wrote to Gordon Wordsworth that the "spoil-heaps" thrown up by quarrying on the Old Man of Coniston were disfiguring the hill. Defenders of the quarries claimed that the heaps would soon weather down and that no new material would be added to them. Collingwood rejected such talk as downright duplicitous. He told Wordsworth that the slag was clearly visible from his own table at Lanehead. Without some measure of regulation, the situation could soon become "serious . . . and if so, the Old Man will be done for."[43]

THE STORM CLOUD OF THE TWENTIETH CENTURY

Collingwood's worries about environmental degradation deepened after the turn of the century. On the night of Edward VII's coronation in August 1902, he witnessed a sinister spectacle. A great low-hanging cloud crawled up the valley of the river Lune from Barrow and Carnforth as far north as Tebay, just ten miles from Kendal. The polluting mist trailed along the flat and low parts of the valley, maintaining a "perfectly continuous" shape, wholly "distinct from the mist of water." On a brilliant winter's day a few months later, Collingwood climbed to the top of Wetherlam above Coniston village and saw the Storm Cloud again. This time, though, it did not halt on the outskirts of the region, but instead "gradually invade[d] the Lake District from the south-east." The phenomenon had a "dun and semi-transparent" body with a "horizontal, clean-cut, upper surface at about 2000 feet." It overtook and fouled other small clouds with a "thick veil." Mesmerized by the eerie spectacle, Collingwood watched the spread of the cloud until it had reached Dungeon Ghyll in the Langdales. When he finally descended Wetherlam by moonlight, he found the valley covered "in a dry, cold fog." He was told that there had been "no sunshine at Coniston that afternoon."[44]

This was not the first time that Collingwood had seen Ruskin's Storm Cloud. Many years earlier, in September 1882, he had spotted it together with Ruskin at St. Cergues looking out toward Lake Geneva. They saw a "brownish grey haze" moving in the fitful wind, veiling the opposite side of the lake "in a persistent thickness of air." A few days later it was still covering the mountains south and east, even blanketing Mont Blanc in gray. Collingwood made two careful sketches of the phenomenon, taking special note of its "smoky, not only thundery, look in the clouds." But he found himself unable to agree with Ruskin's apocalyptic interpretation of the strange weather. Collingwood dismissed the two 1884 lectures on the Storm Cloud as "one of his least convincing though most sincerely meant utterances." In Collingwood's 1893 biography of Ruskin, he interpreted the Storm Cloud in psychological terms only—as a reflection of the master's own turbulent inner states, the product of despair, madness, and fanatical religion, "out of place in these latter days." Yet, just ten years later, Collingwood came to see the Storm Cloud in a new light. The 1903 essay "Ruskin's Old Road," written for the magazine *Good Words*, recounted with great affection Collingwood's trip to Lake Geneva with Ruskin and ventured a material explanation for the observations they had made at that time. The Storm Cloud was the product of coal smoke, "the real enemy of the weather not only in England but in the Alps." "Any one who haunts our Lake district hills knows it well." "You will see it, according to the wind, on either side of Zurich most notably." A strong wind could disperse the Storm Cloud, "but only to deposit it somewhere else." Collingwood's 1902 tourist guide *The Lake Counties* contained several references to this type of atmospheric pollution. Smoke from the furnaces at Barrow soiled the "grass of the mountaintops," blackened the snow, and filled the lakes with scum. In this way, Collingwood came around to vindicate Ruskin's fears, at least in part. Though he rejected the apocalyptic interpretation of the Storm Cloud, he confirmed the empirical basis of Ruskin's observations as "perfectly accurate." The phenomenon of smoke pollution had not been properly "understood twenty years ago," but could now be elucidated through new scientific knowledge.[45]

Collingwood's attack on coal was part of a small but significant movement in favor of smoke abatement among Ruskin's followers. They were responding to the all-too-obvious deterioration of air quality in the metropolis and provinces. Modern scholars estimate that the occurrence of fog in the capital increased almost threefold in the thirty years after 1850. One historian suggests that the concentration of smoke in the London atmosphere peaked around 1890. At the same time, British scientists searched for ways to quantify atmospheric pollution. New methods were pioneered to analyze the chemical composition of the air and measure its sulfur compounds. The second half of the nineteenth century saw a number of initiatives to curb and regulate pollution by activists. Ruskin's friend and ally Octavia Hill helped organize an exhibition on technologies capable of reducing pollution in 1881. This ran for two and a half months in London and was seen by 100,000 people. Afterward the movement coalesced to form the National Smoke Abatement Institution. Ruskin's champion Hardwicke Rawnsley entered the fray in 1890 with his article about Lancashire efforts to reduce pollution. Ruskin's Guild of St. George sponsored the principal of Dalton Hall, John Graham, to write a book on smoke abatement in 1907 entitled *The Destruction of Daylight: A Study in the Smoke Problem*. Graham's work explicitly reinterpreted the Storm Cloud along the lines suggested by Collingwood. He noted that Ruskin's earliest observations of the phenomenon in the 1870s coincided with annual levels of national consumption in the region of 120 million tons of coal. He quoted at length passages on the characteristics of the Storm Cloud from Ruskin, but added to them the chemical analysis of the process of coal combustion and the 1888 calculations of Rollo Russell regarding the social cost of coal use. The latter included an estimate of expenses for cleaning, destruction of textiles, and the loss of work due to poor health and excess mortality. In total, the social cost of coal in London amounted to "£5,470,000, which, divided by the number of inhabitants in the city, came to "about £1 per head." Graham also charted the spread of the Storm Cloud into the northwest region of England. Coal pollution had caused four days of "heavy black fog all over South Lancashire" during

Christmas in 1904, when "people groped, and choked, and slipped in the dark streets." Parts of Graham's book were written in the Lake District, where he had the opportunity to witness firsthand the "dirty brown" haze covering the lower valleys. After discussing the matter with Collingwood, he concluded that the bad air was produced by smoke carried by the wind from distant ironworks and foundries, including those "round the Furness coast." The two Ruskinians also agreed that there was a technical solution to the problem. Collingwood expressed a fervent hope that "modern science" might one day find "a substitute for the crude coal-fire" so that the pollution from Barrow could be eliminated. Graham's book charted in painstaking detail the technological options at hand, expanding on the work of Octavia Hill and Hardwicke Rawnsley in earlier decades.[46]

Beyond this possibility of technical fixes for environmental deterioration, Collingwood saw another reason for cautious optimism. He was convinced that the resilient natural order could heal itself, *given time*. This was a key message in *The Lake Counties*. After all he had seen of Iceland's environmental degradation and moral degeneration, such faith in nature may seem surprising. Yet his optimism was in keeping with his general thoughts on the power of free, self-reliant humans to manage resources in ways that were actually both sensible and profitable in the long run. Like John Ruskin and William Wordsworth, he never associated the protection of nature with absolute absence of human activity. A landscape without the mark of human use, in fact, seems to have dismayed Collingwood. In the letters from Iceland, he had described how he rode his pony through craggy mountainous formations, shuddering at the untamed wilderness. It was "a nightmare of mountains and mist: one huge shape follows another, alike and different . . . till you get quite confused and almost frightened."[47]

For this reason, Collingwood devoted much of his tour guide to the Lake District's long traditions of mining, coppicing, and other economic ventures. Included here were detailed walking plans that featured the landscape's half-submerged industrial past, weaving such descriptions into accounts of natural scenery. As he noted matter-of-factly, "It is difficult to fix a date, but roughly speaking there must have been iron-smelting always going on in the Lake District since the times they were settled by folk civilized enough to need tools and weapons." Collingwood described the industrial ruins of the country with surprising equanimity. Nearly every stream in the Lake District used to have bloomeries (iron furnaces) fueled by great stores of local charcoal: "At one time the woods were nearly destroyed" by "grimy workers" shoveling and hammering away. Yet, despite the heavy use of natural resources, most obvious signs of this busy industry were now gone. This was "a singular instance of nature's *vis medicatrix*, the way she heals old sores." Collingwood drew here on the ancient idea that disease and imbalance could be remedied by natural means: *vis medicatrix naturae*, "the healing power of nature." His pamphlet *The Book of Coniston* from 1897 described a leisurely ramble through a landscape that carried scars of long industrial use—past larch plantations, waterfalls, copper mines, ruined bloomeries, and the trace of ancient glaciers, all jumbled into picturesque harmony. "It is strange to reflect," he mused, "that all of this sylvan wilderness was once a black country." Industrial ruins were as worthwhile to visit as geological wonders. After a brief discussion of local shepherds and a poem about the English Lakes, he described the "sledge road," which was close to "Saddlestones quarry, with its tram-lines and tunneled level, and ... 'rid' or debris." The sight was both unsettling and fascinating—and certainly worthy of consideration. Indeed, glossing over them as if they did not exist might do less good than a matter-of-fact, well-researched commentary. The important part was to keep the impact of industry in the forefront of his readers' minds. No doubt, this embrace of industrial history also had something to do with the particular type of industry in the region. These miners and coppice workers were not automata in a factory line; their "difficult and dangerous work" required training, skill, and perhaps even art: "We can watch the riving of great blocks into slabs and slices, thin as paper in comparison, but sound as a bell." In fact, the rubble and waste left behind by this work included a rich bounty of geological treasures.[48]

Although Hardwicke Rawnsley has long been acknowledged as one of the main forces behind various Lakeland preservationist campaigns, Collingwood may have played an even greater role, albeit one carried out in relative seclusion. Some of Rawnsley's writings seem to lean heavily on Collingwood's far more meticulous scholarship. This may have been a mutually beneficial relationship, since Rawnsley was comfortable in the public eye. Whereas Rawnsley often called attention to the endangerment of nature in the face of industrial activity, his reluctant acceptance of the Thirlmere scheme and his "Sunlight or Smoke?" article indicate that he may have been swayed by Collingwood's more practical position regarding industry and technology.[49]

Collingwood was fairly well convinced that humans could work and play on the land without necessarily destroying it, as long as nature was given time to heal. In fact, the industrial history of a landscape could even prove fascinating in its own right. In the preface to the second edition of *The Lake*

Counties, written shortly before his death in 1932, Collingwood marveled at the changes he had seen in his lifetime: "It is like looking at a kaleidoscope," where the changes wrought were "never for the worse: always into something rich and strange." Motorcars now crowded the roads of the region. Airplanes soared in the skies. Electric power lines extended deep into the Lakes region. Yet this intrusion of modern technology and infrastructure had proven far less destructive to the Lakes than Ruskin had feared. Collingwood did not mention the work of the National Trust in preserving the landscape, but he hinted that there were decent reasons for optimism: "I should like to be one who, as the Romans put it, does not despair of the republic." Later in the book, though, he brushed against a darker thought: what if the mighty forces of social utility and economic progress were to trump the work of wisdom and virtue in the end? He added bitterly: "I suppose you must consider the greatest happiness of the greatest number, as the dogs told the fox."[50]

CHAPTER SIX

Nothing Much

Most mornings, W. G. Collingwood's four children awakened to the sound of Mozart or Mendelssohn played by their mother. They rose and dressed and made their way to a morning room "large enough to contain a small house," with its grand piano and impressive cabinets and tables custom made by an artisan friend of the family. Hung on the walls covered in William Morris wallpaper were their parents' and grandparents' paintings, a large picture of angels by Edward Burne-Jones, and a few drawings by Ruskin. Less important works, abandoned or unsold, rested haphazardly in any available nook or cranny. The smell of oil paint permeated the entire house. Molded plaster festooned the ceiling with its painted motto, *"You do right and God will help."* After breakfast the children filed into their father's study for lessons in Greek, Latin, archaeology, history, literature, astronomy, evolutionary theory, or germ theory. When their work was done, they might set out to climb the Old Man with sketchbooks in hand, or embark on a ramble to mark out ancient pitsteads, geological specimens, and the fauna and flora of the area. Some days, they would take the boat to Peel Island and help their father with his excavation. When it rained, they stayed indoors and mused on their expeditions. They also wrote poetry, fiction, and short essays; they painted and sketched. Then they gathered the best of these materials into a family "magazine," which was sent off to a list of loyal subscribers every month. Over tea and eggs, one of the children would read Ruskin or Kipling to the others, their eyes wandering to the picture window framing the mountains and lake to the west. It drew their gaze down the "steep grass bank, across the lawn, under the fir-trees, through another creaking gate, down the field" to the boathouse as they waited for the sky to clear.[1]

Such was the typical routine of the Collingwood children at Lanehead in the 1890s and the first few years of the new century. When William Collingwood and Edith Isaacs (Dorrie) were planning to marry, they found a cottage available in the Lake District. Collingwood wrote to her that "something caught [him] across the chest" when he heard the news: "I saw my lake and rocks & hills—and I think I began to cry." They leaped at the chance to start a family in the country, where the children could enjoy the full freedom of physical activity surrounded by mountains and rippling lakes. They hoped the children would receive a solid grounding there as well in Ruskin's ideal of sufficient living. A granddaughter, Taqui Altounyan, later wrote in her memoir about visiting the Lake District after World War One: "In our childhood John Ruskin loomed great and mysterious, like a super-ancestor. Our grandparents always talked of him in a special tone of voice, almost as if they were in church." If this was how a grandchild who grew up outside Britain remembered Ruskin, it is easy to imagine what impact he must have had on the children who grew up down the road from Brantwood: Dora (b. 1886), Barbara (b. 1887), Robin (b. 1889), and Ursula (b. 1891). In fact, Ruskin's influence is strikingly evident in the cache of unique materials held by Abbot Hall and the Cumbria Archive Centre in Kendal. These include the family's paintings, stories, travelogues, poems, letters, and vignettes gathered together and "published" in two magazines—*Nothing Much* and *What Ho!*—that circulated for years among friends and relatives. Through painting, storytelling, journalism, natural history, and archaeology, the Collingwood children articulated their own version of the Lakeland Arts and Crafts culture. The magazines reveal a childhood full of simple yet artful amusements, play infused with skill and learning, and always a sense of wonder about the natural world.[2]

By exploring the creative life of the Collingwood children, we stand to gain a better sense of how the ideal of sufficiency became practice. Although Ruskin's own struggles with despair and mental illness hardly invited emulation, Collingwood and his family were fortunate to thrive on his moral precepts. Through them, Ruskin's ideas were passed on to others, such as their family friend, the young, future children's-book author Arthur Ransome. This circle of influence eventually grew beyond anyone's expectations, for it was through Ransome that hundreds of thousands of children across the globe were introduced to the cheerful adventures of *Nothing Much*.

FIGURE 6.1 Left to right: Dora, Barbara, Robin, and Ursula with their mother, Edith Mary "Dorrie" Collingwood, 1894. Photographer: Fred Hollyer, a local photographer in Coniston. Teresa Smith.

"TALKING POSH AND SEEMING STRANGE"

In the late eighteenth century, middle-class parents found new ways of doting on their children. Domestic life offered a welcome reprieve and sanctuary from the strains of the market and capitalist competition. Bourgeois families invested time and energy as never before in their children's education and general well-being. Meanwhile, scholars analyzed the mental and physical attributes of childhood. In particular, the activity of play came under close consideration. It began to signify something more than a state of frivolous pleasure; play was increasingly seen as an essential stepping-stone in both physical and mental development. At the same time, children and the trappings of childhood became an important part of consumer society. Factory production made possible a mass market in cheap toys. Mid-Victorian books and magazines for children initially showed advertisements appealing to adult notions of appropriate gifts, but by the late nineteenth century they began increasingly to target their child readers. The idea of children as consumers in their own right had become prevalent.[3]

The Collingwoods were keenly aware of these prospects and dangers. The family belonged to the upper middle class in cultural terms, with quite substantial connections in the art, literary, and academic world. But their meager livelihood and rural residence insulated them from the main current of urban consumer society. Collingwood and his wife worked constantly, writing novels and painting "potboiler" landscapes and portraits to make ends meet. In pursuing an alternative life in Coniston, they guarded against the corrupting influence of commercial culture, but they seem to have done so with confidence rather than fear. The Collingwoods developed their own activities and their own brand of Ruskinian concerns. This was probably the reason their children did not fit in with the other young people in the village. "[S]mocked, pinafored, and wearing clogs," Dora and Barbara made a few attempts to attend the school in Coniston, but the trial did not turn out well: "the local children hooted at them for talking posh and seeming strange, so it didn't last." The family also employed a "brief succession of governesses." This too proved a short-lived experiment. Yet such awkward interludes did not leave much of a mark on the family. The children's carefully recorded thoughts and adventures show a remarkable mixture of intellectual interests, artistic concerns, and exuberant playfulness. These explorations began in the household and garden but quickly extended outward in widening circles to the local countryside and then the broader world beyond.[4]

The curriculum established by Collingwood for his children at home included books and exercises that united activities both intellectual and artistic, abstract and physical, for Robin as well as the three girls. The children had "lessons in ancient and modern history, illustrated with relief maps in papier-mâché made by boiling down newspapers in a saucepan," in what was very much a demonstration of hands-on learning. The children read widely, of course, including works by Dickens, Wordsworth, Dumas, Kipling, Robert Louis Stevenson, and "their father's passion, the fearful, cold Icelandic sagas." All the children "began Latin at four and Greek at six," but they also learned about the mechanical intricacies of "pumps and locks, oil lamps and water-closets." Collingwood's eclecticism and egalitarianism reflected the ideals of his own teacher.[5] He would have known that Ruskin was a strong supporter of girls' education who believed that girls as well as boys should learn Greek and Latin. Ruskin gave several lectures to the students at the Winnington school for girls, which in turn inspired his Socratic dialogue for "little housewives" entitled *The Ethics of the Dust*. This highly idiosyncratic book was outwardly a series of lectures on crystallography, but Ruskin often portrayed the crystals in moral terms, noting for instance whether they were working as a group or fighting "furiously for their places, losing all shape and honour." The amusing assessment of the resemblance between humankind and the inanimate natural world only went so far, however. Throughout the dialogue, Ruskin steered pupils toward a more complex understanding of the difference between crystals and humans, or dust and soul: "[I]f . . . there be a nobler life in us than in these strangely moving atoms . . . it must be shown . . . in the activity of our hope; not merely by our desire, but by our labour." This was a qualified and thoughtful form of labor, not blind and careless production. The "work and play" the children had engaged in during their lessons should have taught them the true and purposeful capabilities of the human soul, which could be discovered partly through physical activities. In fact, Ruskin encouraged his students to *enact* crystalline structures by standing in bizarre formations together—a quirky lesson, to say the least. The students were told to hold and even break delicate mica in order to understand it better and to inspire further questions. This imaginative but also abstruse book found no favor with the public when it was published in 1864. Collingwood, however, seemed to recognize what genius there was in it, and soon his own daughters were reading it. He must have approved of the progressive mix of mental and physical activity, scientific and moral discussion, and the insistence on both self-restraint and recreation. This was a book that counseled against the desire for diamonds, which only made one "base, frivolous, and miserable." It also claimed that girls' "first virtue is to be intensely happy;—so happy that they don't know what to do with themselves." The business of being happy eventually took on a concrete form in *Nothing Much*.[6]

PLAYING AT SEEING

At first glance, *Nothing Much* might seem like little more than a family scrapbook, containing a loose collection of children's writings and paintings. But a closer look reveals something very different and quite rare in the history of childhood, a concerted effort by four children to make their own monthly magazine, which over the years grew to a high level of intellectual ambition and aesthetic execution, under their father's thoughtful guidance. The first few issues were intermittent, but in January 1897 Dora announced, "The editor of *Nothing Much* begs to inform her readers that the Magazine will now come out ONCE A MONTH. Contributions will be thankfully received. Subscriptions payable to the Editor." The price was listed at "1 d" or one penny. The four authors were precocious, to say the least: Dora was the eldest at eleven, Barbara was ten, Robin eight, and Ursula only six. Subscribers seem to have comprised their extended family and various family friends living both within and outside the Lake District. This circle included about twenty people, plus the six members of the family itself. The magazine was gathered together each month and then sent to the first person on the list; this person would peruse it and then send it to the next person on the list, and so on, until it was returned to the editor. The "published" issue, then, was the *only* copy, and the "subscription" was really a form of manuscript circulation. Whereas other adolescents produced their own newspapers and magazines, thanks to the invention of toy printing presses in the 1860s, *Nothing Much* was a *sui generis* work of skilled labor. The closest model known to the children for this project would probably have been the limited-edition handmade book *Songs of the Spindle & Legends of the Loom*, which was created in large part by Collingwood's friend Albert Fleming. After six years and six volumes of monthly issues, the Collingwood family had amassed a sizeable trove of materials. "[I]n futuredays [*sic*]," Dora wrote, "these volumes will be interesting, both to us & to our descendants." Her father in turn suggested that it might provide inspiration later in life: "You will meet yourselves again, when you're women grown and men,/And you'll laugh to find your masterpieces cribs from 'Nothing Much.'"[7]

Important lessons for life began with vigorous play as the children went rambling, hiking, bicycling, or sailing. For Ruskin, a good education required "active, visual learning." The educational plan for the children of the Guild of St. George workers, devised by Ruskin, included "agricultural schools inland, and naval schools by the sea." The boys should learn "either to ride or sail; the girls to spin, weave, and sew, and . . . cook." Collingwood was not preparing his children for a worker's life, but he placed a similar emphasis on manual skills, opening the door a bit wider for his daughters, who also learned to sail. One of the recurring features of *Nothing Much* was a monthly report on the activities of the Mongoose Club, the purpose of which was to encourage amateur archaeology (usually regarding ancient industry in the region) and the study of natural history while on rambles about the countryside. After a successful expedition, the children would write up a report, call a meeting, and share their findings. Their motto was "Run and find out," very much in the spirit of Ruskin's pedagogy. Members included the Collingwood family along with certain neighbors and visitors. At the meetings they adopted monikers such as Mr. S. D. Fylfot (Dora), Mr. Solomon B. Crystal (Barbara), Mr. Squimpson Ratraps (Ursula—the girls often role-played as men), Mr. Max Howler (Robin), Mr. W. G. Catcher

FIGURE 6.2 Cover of *Nothing Much*, probably painted by EMDC. Abbot Hall.

(their father), and Prof. B. Metaphenelenediamene (perhaps Dr. Cohen, a neighbor). The names perfectly captured the family's blend of scientific inquiry and absurd humor applied to outdoor excursions.[8]

Whimsical wit aside, the children did not lack critical awareness of more serious issues. Dora's account of "The Great Snow" of 1900, as lovely as it is, ends with a description of how the slowly evaporating snow "left black sooty dirt behind, which we suppose was from Barrow smoke." Such pollution right in the heart of the Lake District was a reminder of the pervasive reach of industrial society: so much for Collingwood's effort to raise his family away from cities and factories. Two years later, when he observed smoke snaking its way through the mountains on Coronation night, the matter was debated directly in the club. A paper contributed by the absent Dr. Cohen disputed Collingwood's position: "Perhaps I might be permitted to make a suggestion with regard to the question of soot on the lake, which has been exercising the great minds of 'locals.' According to one authority it comes from Barrow; according to another it travels from the industrial centres on the N.E. [coast]." Dr. Cohen proposed instead that it was simply "Coniston smoke" from the fires lit to combat the chilly mist; the mist had captured the smoke and deposited it on the lake as it fell. Thus, Dr. Cohen set his cozy interpretation of the soot against fears about industrial pollution. By all accounts Collingwood remained unmoved. Indeed, he repeated his hypothesis to the public in his tourist guide *The Lake Counties* the same year. Meanwhile, his children experienced this serious debate in an atmosphere of humor and calm rationality. Their father introduced the children to Ruskin's Storm Cloud without overwhelming them with dire forecasts of the effects of contemporary industries. This tempered perspective is perhaps reflected in a photograph Robin submitted to *Nothing Much* a few years later. The black-and-white image shows the tops of conifer trees and a bright sky being swallowed up by gray, silver-lined clouds. The image, titled "A Storm Cloud," conveys only stark beauty, with no hint of apocalyptic disaster.[9]

Collingwood knew how valuable it was for children to experience a strong connection to the natural world. Many years earlier he penned a slightly maudlin poem about a poor child in the city, who "couldn't stir about:/And through the window was nothing to see,/But bricks, and a waterspout." His own children were able to take in the natural world every hour of the day. Often they engaged in the study of nature, which entailed direct and sustained observation. The stories of *Nothing Much* show the children scouring the woods and hills in search of subjects for stories, paintings, and drawings. The landscape was painted or photographed in all different sorts of weather. Paintings of flowers and studies of birds, feathers, and butterflies abound. Decorative vines adorn margins or frontispieces. Dora's watercolor of a dead chaffinch stands out for its accurate and unsentimental record of nature. Seven-year-old Ursula's crude but lively painting of a siskin is accompanied by the motto "From Nature"—clearly in honor of Ruskin, who strongly advocated drawing from nature directly. Changes in weather and season were followed with great interest; "Flora's Time Table" was a recurring feature that gave the date of the seasonal first appearance of flowers. The children had a formidable knowledge of plants, having grown familiar with "Speed-well," "Ramp and Vetch," "Potentilla," "Milkwort," and "Brooklime," to name just a few. One of the more captivating contributions is a small, thoughtfully executed portrait of "An Unknown Egg" by Barbara Collingwood. A wide margin of blank white canvas frames the diminutive brown speckled egg, with only a half round of shimmering green for a backdrop. She captures its delicate mystery by means of deft shading and a reflective sheen on the apex of the shell. It is accompanied by a description of the location, the nest, the number of eggs, and a bird seen near it. The children say their guide books have failed them, and they turn to the collective knowledge of their readers; their neighbor comes to the rescue—it is the egg of a tree pipit. The sheer quantity of observations suggests a family conscious of the privilege of living in the country.[10]

These images were not just attempts at accurate description and representation. They also aimed to crystallize the experience of nature in words and pictures. When young Ursula recorded her disappointment at missing a meteor shower, she could at least console herself that she had caught the reflection of the moon and three planets in Coniston Water. "[W]e are not lickly [sic] to see it again," she added. The older children captured fleeting moments in time even more self-consciously and effectively. Barbara detailed the children's habit of slinking out into the dusk to see and hear whatever they could. She was eighteen at this time and showed signs of a sophisticated literary taste. "We will go into the Lair and watch it grow dark," she begins. The "Lair" was a favorite nook, a "grassy piece of bank" half-hidden by deliberately placed branches. "You had better bring your cloak," she warns, inviting the reader along. Once in the lair, she gives a swift description of the mountains standing out against the sky, the "pearly" lake, the dark hills, and some "overhanging beech boughs." Visual focus segues to aural attentiveness: "There are a hundred different sounds if we listen for them." She notes the beck and the wind and the new train, "far away . . . above the lake—You hear the whistle before it comes into sight . . . till it rattles faintly into the station." She notes the "boat on the lake, splashing in & out among the black stillnesses," and commands her readers to "Listen to that shriek up in the wood . . ." It might be an owl, she says, but "I think it is some spirit." Shivering, they emerge from the lair and cover

FIGURE 6.3 Painting of a dead bird, 1898, by Dora Collingwood. Abbot Hall.

it back up for another night. With its fine detail and intimate use of second person and present tense, the piece captures the moment for the family, for the children's future selves, and perhaps for their future children. By the end of the vignette, readers may well feel they have been inducted into the mysteries of Coniston at night. The piece is more than a mere observation of nature; landscape here is self-consciously filtered through the lens of individual experience. Ruskin had emphasized the importance of learning to *see*, not in order to become a better painter, but in order to appreciate nature and to enhance and develop one's intellectual and moral faculties: "Landscape can only be enjoyed by cultivated persons; and it is only by music, literature, and painting, that cultivation can be given." In Ruskin's view, the ultimate goal of education was not merely to produce talented young ladies and gentlemen, but also to teach them to appreciate and safeguard landscapes for the future benefit of the nation. Landscape was to be experienced as a *"memorial,"* a place whose true importance was accessible only to "cultivated" people able to appreciate the "sacredness of landmark that none may remove, and of wave that none may pollute." "[A] nation," he argued, "is only worthy of the soil, and the scenes that it has inherited, when, by all its acts and arts, it is making them more lovely for its children."[11]

Natural history was never divorced from human history for Collingwood and his children. *Nothing Much* encouraged the children to think broadly about the physical, tactile reality of the landscape, including the coppice woods, active quarries, abandoned buildings, and archaeological sites. The industrial history of the Lakes formed a common theme of Mongoose Club expeditions into the surrounding hills and fell lands. On a ramble near Brantwood, Dora and Robin set out "one frosty morning" equipped "with compasses and appliances for surveying." They came upon a "midden-stead" or ancient trash heap—places often ripe for excavation—as well as "a singular growth, much resembling a lizard, on a cherry-tree" (sketched on the last page). Finally, they reached a pitstead "entirely covered with copse, thereby showing that it had been out of use for a long time." Coppice wood had once been burnt into charcoal at these old pitsteads. The children were thus inspecting the vestiges of a local industry that had been practiced for centuries. During their explorations of the Coniston countryside, they gradually uncovered the history of the landscape, acquiring a sense of how nature recovered from human impact over time; they noticed the manner in which natural resources—the trees themselves—had been harvested systematically but not destroyed. There was no sense of absolute competition between humans and nature here; nature *could* be used wisely.[12]

The Mongoose Club findings were presented at their meetings in the form of mock-serious antiquarian articles. In 1901 Robin began examining a particular site "above Atkinson Ground," which he proposed was an old watermill. His father was skeptical, suggesting that it might just as well be the remains of a house for "domestic fowl." The club decided that more information was needed. Robin revisited the site and delivered a second paper. This time he had also discovered "2 pieces of wall . . . which seems to have the remains of an aqueduct above it." Still, their father was not entirely convinced. On the third visit, they found "slates and rafters," indicating it had a roof, but Collingwood hesitated to call it a "water-course." The amount of time and energy spent assessing this potential site of human industry was typical. Collingwood's novel *Thorstein of the Mere* had explored the deeper roots of human settlement in the Lakes. The Mongoose Club often compared their father's fiction with the material remains in the landscape. Their father ("Mr. [Wild Goose] Catcher") praised their efforts to observe and describe, but also grilled them on their findings. Had they spotted the "small triangular holes in the rock under New Bridge at Greenodd"? These he believed had once been used for casting arrowheads by men like Thorstein, more than a thousand years earlier.[13]

The family's fascination with archaeology did not preclude a gleeful interest in the wonders of modern technology as well. The children were hardly Luddites. Barbara came across men at the quarry skillfully splitting slate, while she was collecting "fossils & luckystones" and scraps of Coniston limestone. Nor did they shy away from decidedly destructive activities. They loved visiting the shooting range enough that Barbara complained, "we've been there so often, & never find any bullets now." Target shooting was just one more activity—not perceived as necessarily threatening the natural world. Robin pasted in drawings of dreadnoughts into one issue of the magazine. Dora submitted an enthusiastic Mongoose Club account of "The Big Blast" after a local slate quarry dislodged "40,000 tons of rock" by means of explosives. The entire affair was cast as a grand excursion into nature. After traversing "a little wood full of thickly blossoming rosebushes," the children came across "a fernfringed beck" from which to drink. Dora observed the way the vegetation grew over "disused workings" and remarked that "the quarries hardly disfigure the valley," since nature could recover the land when humans were done with it. Finally, she described the explosion with no attempt to conceal her delight: "Bang, boom, boom-m." There is an echo here of their father's views: a healthy natural environment could coexist with human industry without devastating the landscape long-term. Disruptions to the natural world were temporary and confined to specific working areas. In other words, we should perhaps not make too much of Dora's enthusiasm for explosions. Had the quarries laid waste to the hillside with gargantuan mounds of rubble and spewed sooty smoke across the lake and Lanehead's garden, she would probably have felt differently about the whole affair.[14]

MAKING AND MOCKING

The children were forced to compromise and adapt critically with regard to the world of advertising and consumption. By reading newspapers and magazines, doing the shopping in Coniston village with their parents, or taking trips into the cities, the children were exposed to all kinds of advertising materials and commercialism. Ruskin had refused to advertise his own publications, objecting "even to the very small minority of advertisements which are approximately true." Collingwood sometimes seems to have balked at the idea of promoting his work by means of advertising, though he was a great deal more pragmatic than Ruskin on this score. He wrote to an editor friend about the guidelines for a new journal dedicated to Ruskin and the Arts and Crafts movement, saying that advertisements were "entirely anti-Ruskinian." Nonetheless, he noted frankly, "I don't object an atom to them."[15]

The Collingwoods raised their children to resist the lure of commercial jingles, catchphrases, and announcements. They were gratified to see them develop a critical stance toward the consumer world. Advertisements became objects of study that they could play with, admire, transform, or reject wholesale. For example, Robin carefully cut from newspapers a series of advertisements for cocoa and presented them in one issue of *Nothing Much*. At the start of the series, he produced his own title: "'—Is The Best.' No. 1. 'Cocoa.'" The dash signaled a blank space, a deliberate omission of one word, namely, the particular brand of cocoa. This was followed by seven cocoa advertisements featuring wildly superlative and often bizarre praise for the products. It is not certain what Robin meant by bringing together these clippings, but it seems a fair guess that he questioned how they could *all* be "the best." Lining them up one next to the other certainly took the punch out of their claims. These included:

> van Flouten's Cocoa: "The Best in Existence—the Cheapest in Use"
> Cadbury's Cocoa: Absolutely Pure. Therefore Best.
> Dr. Tibble's Vi-Cocoa: "Something Better Than Others" . . . "refreshing and agreeable, whilst wholly free from the objection of being either sickly or insipid"
> Suchard's Cocoa: ". . . beyond competition"
> Dunn's Cocoa: "'so perfect, so peerless'—Shakespeare"
> Epps's Cocoa: "The Most Nutritious . . . Grateful—Comforting"
> Fry's Cocoa: "'No Better Food.'—Dr. Andrew Wilson, F.R.S.E., &c."

Most of the advertisements featured boldfaced fonts and images of people enjoying cocoa. The purveyors of these cocoas may have found success using such slogans in general, but in this case Robin's shrewd compilation helped the reader take on the point of view of both the seller and the buyer. A few years earlier, Dora had sketched what may have been a mock-professional advertisement of her own for "Locklick's Cocoa" featuring a smartly dressed serving maid carrying a tray of steaming cocoa cups and the unbeatable sales pitch "at all grocers." Later the same year, she included a sketch of a cake for "Kirkbride's Wedding Cakes." It is unclear whether she copied these from a newspaper or came up with ideas for an imaginary or existing brand. However, their presence in the magazine suggests something of an insider's understanding of the commercial manipulation of desires.[16]

FIGURE 6.4 Cocoa advertisements compiled by Robin Collingwood, aged 11, for the March 1900 edition of *Nothing Much*. Teresa Smith / Abbot Hall.

A critique of advertising made it into the magazine in other ways, too. One of the children sketched a scene showing two boys plastering a "Happy Xmas" sign over political posters and public announcements (mostly obscured). The scene conveys an absurd and hopeless competition for visual space. The December 1899 issue of *Nothing Much* featured a longer piece on the Queen's visit to Bristol, which exuded a complicated mix of excitement, awe, and satire. The anonymous contributor (listed only as "the special correspondent") also displayed a critical attitude when describing the merchants' preparations for her reception. A fishmonger dragged a pew from a neighboring chapel and set it on his marble countertop, evidently to get a good view. Elsewhere "the 'stores' announced 'We love our Queen'; a jeweler offered 'A "Brilliant" Welcome to our "Diamond" Queen.'" The correspondent noted that the "Welcome Victoria" banners were put up "in a manner seemly & becoming." Yet this must have been said in jest, because he or she also included an illustration showing their jarring juxtaposition with advertisements for Quaker Oats and Bovril. It was a good-humored jab at how the royal visit became an occasion for the hawking of cheap wares. The report was meant to amuse the entire family.[17]

This is not to say the Collingwoods forbade their children toys or other

FIGURE 6.5 Cartoon showing two boys pasting a Happy Xmas sign over advertisements. *Nothing Much*, October 1899. Abbot Hall.

goods. A letter to Dorrie in 1894 included Collingwood's partial list of presents picked up in the city: "I am sending also a printing press to Robin ... Also a cannon ... I have bought a good paint box for Dora, also a race-game of the [Natl Western line]." Four years later, the children were treated to a Christmas tea at the home of the Marshalls nearby. Barbara described the beautiful decorations and several gifts set out for each of them. Robin received another "big printing press," and the girls got chocolates, "basket chairs," books, and a doll. One of Barbara's paintings showed a little girl with her mother at a doll store; the colors are warm, and the scene is full of childish pleasure. "Ursula and Robin are playing shops with clotheshorses & dust-sheets over them," Dora wrote her mother in 1898. This signaled a healthy fascination with the consumer experience, once again from the point of view of both buyer and seller. But imaginative games or learning skills of the hand and eye took up most of the children's time. They planted their own garden, devised an elevator basket to lift kittens in and out of a window, and practiced bookbinding. On the whole, *Nothing Much* (which was always made by hand, not printed in multiple copies, despite Robin's presses) focused on discoveries relating to nature, archaeology, art, fiction, and poetry, all outside the sphere of commerce.[18]

SHARING WITH BARBARIANS

Long before the advent of *Nothing Much*, in the summer of 1887 Collingwood had taken his wife and (a very young) Dora for a rough, monthlong mountain stay. With a hint of horror, Susanna Beever relayed the news to Albert Fleming: "The Collingwoods [are coming] into a cottage at the mines to-morrow — they will have very primitive doings — & how will they cook?" The Collingwoods employed a housemaid and a cook in their home, but judging from Beever's account, they rather enjoyed doing without when they were camped near the head of the mines halfway up the Old Man and Weatherlam. Likewise, there were plenty of opportunities for roughing it on small, rocky Peel Island in Coniston Water, a place that enthralled the family for at least three generations. Barbara's account of one expedition suggests that much of its draw was the way activities too often taken for granted were suddenly rusticated and therefore more satisfying. One day, they made their way by boat and tried to light a little fire. After a long struggle with the wind, smoke, and an overturned kettle, and having been forced to start all over again, they finally enjoyed their tea. The long delay enhanced their gratification: "[I]t was good when we got it." She explained,

FIGURE 6.6 Peel Island, Coniston, a site of archaeological excavation and "rough" outdoor teas.

"[t]he advantage of tea at Peel Island is that the jam seems to be thicker on the bread there. I think it's the atmosphere, and the bluebells." It is no wonder, then, that Dora later carefully copied out for their magazine a saying her father was fond of: "Why are Frenchmen said to be very abstemious? Because they say 'Un oeuf is as good as a feast'" (with a bilingual pun on *enough*). In all these examples, the children's understanding of the ideal of sufficiency appears at the same time playful and mature, rooted in an appreciation for the natural world.[19]

The fame of the Lake District was connected with water, sailing, rambling, and picnicking. But the success of tourism also threatened to diminish the charms of the region. The stream of visitors into the Lakes was growing by the year. Collingwood faced a double bind. If Ruskin's ideals were to be anything more than the preserve of a tiny elite, then the Lakes and other beautiful landscapes would have to accommodate increasing numbers of visitors. Collingwood struggled with this issue in a number of poems included in *Nothing Much*. In "The Invasion of the Barbarians," he bristled at the "careless, curious eyes from far and wide" and wrestled with the sense that they were overtaking and demeaning something that was *his* by dint of habitation and knowledge: "my lake—all mine by right of choice—made common to the trippers' crowd." He was peeved by the way they spoiled the rural beauty with their noisy roughhousing. Many of the tourists seemed to Collingwood to lack the appropriate care and appreciation needed to enjoy the Lakes. Yet he eschewed simple snobbery, musing uneasily, "Their choice being mine, why should I not be proud?" All the same, the threat of population pressure to the fragile beauty of the land remained—as did the fear that merely vacationing in the Lakes was fundamentally distinct from appreciating it in the manner Ruskin had advised.[20]

In another poem Collingwood took on the perspective of a traveler in Germany, as he grappled further with the problem of tourism and industrialization abroad. Here, he lamented the increasing pollution as well as the loss of Old World splendor. He wondered what was left to see "In this— their twentieth century?" Evidently, instead of castles, barons, or "crag and vine," they saw "lumbering clouds of smoke." He also mocked what he saw as Germans' tendency toward excessive consumption in "Supper at Mainz," a clever piece of doggerel. In this poem he draws a firm line between the British and the Germans with their "endless dinners":

> The Germans have their eel, eel, eel,
> Then a chunk of veal, veal, veal,
> Then a squash of vegetable
> Then some chicken, if you're able.
> Pudding, cheese, and peel, peel, peel
> Fruit to end your meal.

There is little surprise that "it makes 'em fat, fat, fat!" Through all the foolish, rhythmic chanting, the sermon on sufficiency and self-restraint resounds. He concludes the poem, "I munch my modest ration/in the manner of my nation." However, his pronouncement on his nation's superiority sounds a little bit like wishful thinking. But one can imagine that the silly, malicious wit of this poem took root in his children's minds.[21]

ORDERING THE IMAGINATION

The energy expended on the production of a family magazine distributed to roughly twenty people might seem inordinate to some, yet for Collingwood the end of the project easily justified all this labor. *Nothing Much* was the playground and workshop where his children could develop their imaginative faculty to its fullest extent. This was also a cardinal feature of Ruskin's ethics. For Ruskin, the freedom and happiness of the medieval artisan found expression in the aesthetic forms of the Gothic style. In the modern world, the link between work and art had been severed, banishing the play of the imagination from labor and the production process. At the same time, commercial society encouraged a continuous expansion and fluctuation of human wants. Both producers and consumers were constantly imagining new goods and new forms of satisfaction. In his critique of liberal political economy, Ruskin proposed that it was possible to reorient the imagination away from the marketplace, toward the world of art and nature. If the imagination could find other objects of desire, shaped by knowledge of the natural world and guided by artistic skill, this would drastically undermine the power and appeal of commodities.[22]

The Collingwood children were in fact encouraged to exercise their imagination to an extraordinary degree. Whole worlds of new creatures were invented in the pages of *Nothing Much*. The monthly deadlines and Mongoose Club meetings helped keep constant pressure on the children to expand their sense of fiction and fantasy. Stories of imaginary lands, detectives, and adventurous animals were standing features, and were often developed over several issues of the magazine. Robin invented the ornithology and ecology of Jipandland in careful sketches and precise classification. The "Paw Hawk," he tells us, "inhabits the rocky Harvoëcio" and "feeds on the Redtail owl, and on Pothigs." His birds were the product of *seeing* nature according to Ruskin's dictum, but also of creating lifelike fictitious beings, virtual toys, in a manner of speaking. A similar effort went into Dora's beautifully illustrated tale of "The Cats' Christmas Party." Set in an abandoned pitstead and vaguely reminiscent of *Alice in Wonderland*, it was at once earthy and magical. Ruskin argued that children ought to learn "morality" by being "taught gentleness to all brute creatures." Because the Collingwoods were indebted to their neighbor Susanna Beever, her insistence on kindness to animals likely influenced them as well. There were perhaps few better ways to achieve this than to engage in a bit of humorous anthropomorphizing. In another standing feature, Robin and Ursula explored "The Customs of Cat Land." This set of articles gave an ethnographic account of the origins of the "feline race," noting that human "fires and hearthrugs" first drew them indoors in "a great invasion." They went on to detail these cats' "Clothing, Eating, Education, Amusements, etc.," in different chapters with descriptions and drawings. The "clothing" of the cats borrowed elements from Sikh traditional attire; their "Education" offered track-and-field exercises;

FIGURE 6.7 Painting that accompanied Dora Collingwood's story "The Cats' Christmas Party" from *Nothing Much*, September 1898. Abbot Hall.

feline combat resembled medieval warfare with knightly hierarchical structures. Culinary tradition was given pride of place: "Eating plays an important part in Catland . . . a good cook is always respected." The children learned about other animals as well. *Nothing Much* featured notes on frogs, poultry, and even a three-legged lamb. The children's beautifully crafted magazine entries, so often inspired by animals, conjured up a whimsical connection to the landscape.[23]

The children's pleasure in the natural world appears in a completely different sort of activity as well—sailing, a skill Ruskin encouraged children to learn. Boats were common enough on the lake that they could easily have been taken simply as a mundane form of transportation, or perhaps as the privilege of careless leisure. But the Collingwoods felt that sailing was much more than a physical act. Paintings, stories, poems, and photographs depicting boats graced dozens of pages of *Nothing Much*. If properly learned, sailing could align the human spirit with the elements. As with their mountain excursions, real danger often accompanied beauty and exhilaration. Some of the Collingwoods' stories featured vacationers at risk of drowning in the lake. In fact, few activities marked a person's status more decisively in the family lore than whether they knew their way around a boat. In Robin's "Two and a Night Line," the main character rescues a charming woman on holiday who has foolishly fallen overboard. She later breaks his heart, more or less proving her unworthiness on all counts. Collingwood wrote another story called "Things you must NOT do in a boat," illustrated by Barbara. Here again, a pretty woman seems not to understand safety protocols. Though she has "been in boats lots of times," her companion suspects she has only been "with trippers" (i.e., day-trippers), and does not really know anything at all. The essence of sailing came down to skill and knowledge rather than wealth or leisure. One had to train and be alert or suffer the consequences; the proper understanding of wind and water did not come easily. Tourists were missing out. These stories explored the idea of sailing as an artful activity, while also providing a guarded critique of vacationers.[24]

The sailboat offered a powerful metaphor for the well-ordered life. Collingwood had learned this years earlier when he and Alexander Wedderburn translated Xenophon's *Economist* for Ruskin. Socrates relates how a wise gentleman-farmer named Ischomachus once described to him an impressively organized Phoenician merchant ship. There, in a relatively compact space inside the hold, all the necessary tools and food and goods were arranged so carefully that everything was always "quite ready to hand." There would be no time to run around looking for things in the middle of a sudden tempest. One had to plan ahead to ensure safe sailing, considering in advance where to put things. Ischomachus noted that the same principle should be applied in the home, which after all had far more storage space than a ship. Whether a house was large or small, the important point was careful organization. Ruskin and Collingwood saw in the ancient Greek treatise a compelling philosophical definition of wealth in terms of order, usefulness, self-restraint, and the satisfaction of simple needs. The image of the orderly boat was also an apt metaphor for the happy soul.[25]

In this vein Collingwood wrote "The Launch of the Dora" for his wife when their daughter turned fifteen and left for a long stay in Ternitz. In the poem the vessel is built "taut and trim, on even keel,/True, and yet not too stiff to feel." She must travel far, to "alien shores," but he imagines that one day she will come back on "kind breezes." Her imagination and self-reliance would have to see her through. Many years later, in 1916, Collingwood told Dora that he wished he could have done more for her, but at least she had not had "a dull or stupid life" and she had "made use of it." She had just married a friend of Robin's from college, an Armenian physician, and begun a fascinating and complicated new life in Syria. She came home on short visits now and then over the years, but she would not return for good until 1957. Such a life voyage certainly required a soundly built boat.[26]

RUSKIN'S GRANDCHILDREN

It is tempting to look to the Collingwood children's later lives for proof that *Nothing Much* and other activities shaped their careers and households in a Ruskinian fashion. But the line of influence is not straightforward. Ursula became a midwife, and the other children took on artistic or intellectual work. Dora painted; Barbara took up sculpting; and Robin became an Oxford don and philosopher of history. Busy with family and careers, neither landscape preservation nor social reform seems to have taken up a lot of their time. The vagaries of modern politics probably had something to do with this. The outbreak of war in 1914 opened up a Pandora's box of hopes and horrors, just as Collingwood's son and daughters entered adult life. The relative equipoise and serenity of Victorian society was succeeded by an age of extremes. The slaughter in the fields of Flanders eclipsed even the worst nightmare visions of the Storm Cloud. Their parents' enthusiasm and noble intentions could hardly prevail against such forces. In this Brave New World, the gentle idealism of the Arts and Crafts movement faded from the scene. Going against the tide, Robin Collingwood praised the sage of Brantwood as "the best-equipped mind of his generation" and "more independent and original" than his contemporaries. But his final work of political philosophy from 1942, *The New Leviathan*, was occupied with the threat of Nazism and fascism to the liberal state, not the revival of guild manufacture or the practice of ethical consumption.[27]

Yet there is more to the story. When we turn to the third generation—W. G. Collingwood's grandchildren—Lanehead found a new lease on life in the most unexpected way. Dora's husband, Ernest Altounyan, who had trained in medicine at Cambridge, took her and their first child to start their lives in Aleppo, where his father ran a medical clinic. They had four more children over the years. Dora's eldest daughter, Taqui Altounyan, wrote a lovely memoir of life in Aleppo with extraordinarily vivid accounts of mountain vacations, sailing adventures, and a Scouts club. Dora later decided to send her children to boarding schools in the Lake District and Peak District. Despite Taqui's love of Syria, life at Lanehead cast a "magic spell" on her. Long afterward she would remember how her feet crunched on the gravel as she made her way across the "quarried slate threshold" into the house. Like their parents, the Altounyan children fell in love with Peel Island. They found the old oars and sails of the boat "all covered with cobwebs," but soon made their way out to the lake. They stayed overnight on the island, where they made a fire, but could not bring themselves to use the spirit lamp: "'You can't do that on Peel Island.'" Taqui's gift for descriptions of nature and practical skills like sailing seems to spring from the same impulse that guides *Nothing Much*, though she says nothing of Ruskin's philosophy. Yet something of their mother's (Dora's) world seems to have been handed down to them, and in time it made its way into a brand-new work of fiction.[28]

The world of *Nothing Much* found a novel expression in the stories of family friend Arthur Ransome. Ransome met the Collingwood family one summer's day in 1903. Coming down the Coppermines Beck from the hills, William Gershom found the inert body of a young man on the side of the stream. He was greatly relieved when the "corpse lifted its head" and told him that he had "been trying to write poetry." They walked back to Coniston together in friendly conversation. Ransome was surprised to discover that the older man found the writing of verse a "reasonable occupation." At the time Ransome was an aspiring writer and apprentice in the small London publishing firm of Unicorn Press. When they parted ways in the village, Collingwood introduced himself and Ransome realized that this was the author of *Thorstein of the Mere*, "the best loved book of [his] boyhood." He also remembered that he had met Collingwood before on Peel Island during a family picnic. A few days later Ransome summoned all his courage and went on a visit to Lanehead, where he met the whole family.[29]

This encounter left the young poet thunderstruck: "Nothing that I can write can adequately express what I owe to W. G. Collingwood and his wife." Ransome had lost his father early in life and felt a lack of confidence and direction. But Collingwood and Dorrie, along with their children, supplied him with all the love and attention he needed, and a great deal more besides. The following summer, when Ransome returned to Coniston, Dorrie told him to give up his lodgings at the Waterhead Hotel to come live with them instead. From this moment onward, he was "adopted" by the family. The rest of the vacation went by in "a golden haze." Young Ransome was enchanted by Dorrie's piano playing, the good cheer of the family, and the sense of collective work: Collingwood in his study and the three daughters, now teenagers, "at work with easels or with clay in the Mausoleum, a tumble-down conservatory." At lunch they went "forth with bun-loaf and kettle" to talk and read for a while in the open air. He appreciated the Collingwoods' artistic love of work for its own sake: "At Lanehead work and not its material rewards was the only thing that mattered."[30]

Despite his good fortune, Ransome found it difficult to believe that this was now *his* family and his life. The entire situation possessed an otherworldly beauty for him. One afternoon Mrs. Collingwood entrusted him with money to make purchases on her behalf at a shop across the lake. Near the hotel grounds he noticed "a hawthorn tree that had shed its petals all about it, a patch of glittering snow on the dust." The tree seemed a mysterious sign of everything that he had gained in his chance meeting with Collingwood. Years later he observed in his autobiography: "During all that time the leaves of the trees seemed more luminous than they are today and the hills had sharper edges." He stood there "gaping at this or that as if I feared I should not remember it for ever." Long before he met the Collingwoods, he had possessed this power of mental concentration and reverie. His earliest memories had the same startling intensity. But now Ransome's sensibility was imbued with a broader ethical ideal. This was in fact a very Ruskinian activity: looking closely at nature with one's most intent gaze. The keen sensibility cultivated at Lanehead showed Ransome that living well required finely trained senses and a wide variety of skills.[31]

Twenty-five years later, Ransome became the chronicler of just this sort of life in his beloved books for children, starting with *Swallows and Amazons* (published in 1930). The "golden haze" of Lanehead is lovingly re-created in these stories of Picts, pirates, and polar exploration, set in a landscape that combines features of Coniston and Windermere: Peel Island appears as Wild Cat Island, Lanehead is Beckfoot Farm, and the Old Man of Coniston is Kanchenjunga. Much of the narrative centers on surprisingly technical descriptions of manual skills—sailing, camping, boatbuilding, and charcoal burning—the likes of which had seldom been seen in children's literature before. Ransome's inspiration can be traced back to his experiences with the Collingwood family and then to Dora's children in particular—the characters in the stories have their names (except for Taqui). In the summer of

1928, Ransome had even taught the Altounyans how to sail on Coniston Water in a stout dinghy named *Swallow*. Ransome's talent was to translate the appeal of the simple life into a concrete language that children could understand. This had been Collingwood's intention as well in *Thorstein of the Mere*, but it was Ransome who found popular success. Through *Swallows and Amazons* and its sequels (including films and TV adaptations), generations of English children discovered the charms of Lanehead and a distant echo of Ruskin's thought without knowing its origin. Ransome's chance encounter with Collingwood at the Coppermines Beck thus gave Ruskin's political economy and moral philosophy a secret new life.[32]

This line of inheritance from Ruskin through W. G. Collingwood and the children of *Nothing Much* to Ransome was also a process of forgetting. A love of natural history, manual skill, art, and rustic life bound them all together, but the motives and worldview changed dramatically over the generations. Ruskin's Storm Cloud vanished first. The threat of environmental degradation was still palpable for W. G. Collingwood, but had lost all urgency with Ransome. There was no fear of industrial blight or smoke pollution in his stories. At most he included some brief asides between the children about how it was wrong to litter. After *Nothing Much*, Ruskin's systematic critique of consumer society faded away. There was no trace of Brantwood and its Professor in Ransome's books. What remained was a vacation version of Ruskin's ethics. Ransome's readers explored the Lakes while on holiday from the city. The skills they learned, *if* they learned them, were of little use in everyday life. Their love of the country and sailing were not part of any radical critique of urban life and industrial production, only a joyous but temporary suspension of reality.[33]

These changes were already underway in Collingwood's work. The tourist guides he wrote about Coniston and the Lake Counties sought to diffuse some knowledge of the landscape and its inhabitants to day-trippers and holidaymakers, but there was no pretense that they should be persuaded to drop out of society and join the community as statesmen-farmers or painters. It is both ironic and understandable that the success of the National Trust drove this line of thinking to its logical conclusion, as land use and commercial interests were regulated in order to preserve the landscape and bring in tourists. The success of Ransome's stories bolstered this tendency. While the protection of the Lake District has been a wonderful achievement, in the larger scheme of things, it is only a step toward a much more difficult goal. The great question now is whether the ideal of sufficiency will become a dominant principle in modern culture, as Ruskin once intended.

CONCLUSION

Ruskin in the Anthropocene

On the hill above Brantwood, up the Zig-Zaggy Path through the coppice woods, Ruskin found the stationary state in the winter of 1875. His walks led him regularly to the edge of the forest, and the stone cottage of Lawson Park, where he befriended the nine-year-old girl Agnes Stalker and her family, a shepherd with his wife and their eleven children. The cottage seemed to him more or less exactly what he wanted for the tenants of the Guild of St. George. The modest parlor, kitchen, and bedrooms of the house would easily fit in the "average dining room in Grosvenor Place or Park Lane." In an early sketch for the guild, Ruskin explained that he encouraged its members to cultivate the land "with their own hands, and such help of force as they can find in wind and wave." The factory system had no place in this economy: "We will try to make some small piece of English ground, beautiful, peaceful, and fruitful." He added: "We will have no steam engines upon it, and no railroads." In other words, the guild members would act as if England had run out of coal and the mineral energy economy had come to an end. This was Ruskin's version of the stationary state, the stage of history many Victorians believed would come about when economic development confronted the physical limits of growth. Because of their poverty and isolation, Agnes Stalker and her family already lived as Ruskin intended his guild members to do. Lawson Park lay fifteen miles from the closest manufacturing town. The children of the cottage saw "no vice, except perhaps in the village on Sunday afternoons." Agnes knew the "mysteries of butter-making and poultry-keeping" but little or nothing of city life and modern technology. In *Fors Clavigera*, Ruskin explained how husbandry without machinery could become the basis of a happy life: "A man and a woman, with their children, properly trained, are able easily to cultivate as much ground as will feed them; to build as much wall and roof as will lodge them, and to spin and weave as much cloth as will clothe them." Ruskin thought this simple life would endure long after the great structures of industrial society had vanished. One day the embankments of the railways would be "ploughed down again, like the camps of Rome, into our English fields."[1]

Ruskin's visit to Lawson Park suggested to him the idea of the *Bibliotheca Pastorum*—a library tailored for "British peasants." On a whim, Ruskin asked Agnes whether she had received any books for Christmas. In response, Agnes brought him her whole library—a "good pound's weight" of woodcuts and cheap penny prints, "in the best Kensington style." Ruskin carried one of these penny pamphlets back to Brantwood, where he read it with disgust and sadness. He found it stupidly sentimental and vindictive, full of empty religion and vacuous morals for spoiled middle-class children. But what kind of education would be more appropriate for Agnes? To foster her "secular virtues," Ruskin suggested a combination of practical husbandry and natural history. The guild ought to give Agnes "a yard or two square" of ground, "which should be wholly her own," together with tools and seed and a beehive. For her "crowning achievement," Agnes would be asked to "produce, in its season, a piece of snowy and well-filled comb." She would also learn about bee behavior from a few significant books of natural history. In this way the hive provided a lesson in the moral authority of nature: bees were orderly, industrious, and cooperative, but only if their little society was well managed by a good steward.[2]

The next step in Agnes Stalker's education was a reading course in the classics. Ruskin chose Xenophon's *Economist*—in the translation by W. G. Collingwood and Alexander Wedderburn—to be the first title in the *Bibliotheca Pastorum*. This ancient Greek treatise on good husbandry and household self-sufficiency demonstrated the "unchanging truth" of a society rooted in agriculture. The proper aim of political economy was to produce adequate subsistence from the land, not to amass huge industrial fortunes for a few while confining the poor in factory slums. Agnes and her family were better off in their tiny cottage than the affluent middle class in their suburban terraces. Much like his nemesis the liberal economist John Stuart Mill, Ruskin thought of the stationary state as a moral choice rather than an economic necessity. Ruskin never doubted the power of humans to command nature in order to create artificial worlds. He happened to be traveling through the manufacturing districts of Northern England as he pondered the problem of Agnes's education. In Bradford he saw a sprawling new suburb on the southeast side of town—one of the most "frightful things" he had "ever yet seen." The growth of the factory town filled him with a fearful kind of "reverence" at the "infinite mechanical ingenuity of the great centers." Such "fierce courage and industry" was more like the "fervid labors of a wasp's nest," which at "the end of all is only a noxious lump of clay." In contrast, Agnes's beehive offered a miniature lesson in moral land use: how to harvest natural wealth rather than feed the Storm Cloud with coal. But such high hopes could not insulate Agnes from the cruel pressures of rural poverty. When Ruskin visited the cottage again a few years later, he found a new family living there. Agnes and her sister had been sent into service after their mother died.[3]

The story of Agnes Stalker illustrates very well the driving question of Ruskin's social vision. His early concern in *Modern Painters* and *The Stones of Venice* with aesthetic judgment and the beauty of Gothic architecture drove him to consider how art was linked to the social condition of labor. In later works like *Unto This Last* and *Fors Clavigera*, this fascination matured into a theory of the good life, focused on the virtues of skilled labor, the limits to growth, and a rejection of industrial mass production. Ruskin's followers in Lakeland inherited his utopian vision and turned it into a practical culture of sufficiency. The revival of handicrafts, the founding of the National Trust, Beever's gardening, the battle over the dam at Thirlmere, and the Viking historical fiction of Collingwood were all attempts to formulate a viable alternative to modern consumer society. Some projects were aimed specifically at the rural poor, like Fleming's linen manufacture and Ruskin's *Bibliotheca Pastorum*. Others brought the question of consumption to bear on the everyday life of the middle class, whether in gardening, preservation, or children's education. Common to them all was the urge to channel human desire into progressive activities connected to the natural world and the agrarian traditions of the past.

In twentieth-century Britain, Ruskin's reputation underwent an eclipse. The massive economic growth of postwar affluent societies—what some historians call the Great Acceleration—seemed to relegate the Victorian idea of the stationary state to the dustheap of history. However, the furious pace of growth also revived long-standing worries about the physical limits and costs of economic development. Indeed, the accumulating evidence of global warming in recent decades greatly strengthens the environmentalist case against modern consumer society. We appear to have entered a new geological epoch—the Anthropocene—when the human species for the first time has become a geophysical force in its own right. The carbon dioxide emissions from our fossil fuel energy economy are rapidly changing the physical conditions of the biosphere with rising sea levels, acidification of the oceans, mass extinction of species, and extreme weather. There is a real danger of hitting a tipping point of catastrophic warming in the near future due to positive feedback mechanisms (when small initial changes give rise to a self-reinforcing and accelerating pattern).[4]

The arrival of the Anthropocene rekindles the deep worries about mass consumption and industrial production that haunted Ruskin. Modern notions of freedom, democracy, and equality assume ever-increasing standards of living. But the material forces of modern consumption—our power to tap into fossil fuel energy to make things cheaply on an unprecedented scale—now threaten to undermine the basic life-supporting capacity of the biosphere. The horror of Ruskin's Storm Cloud—a sweeping tide of pollution and climate change that transforms all of nature irreversibly—is beginning to look more like an eerie premonition of things to come than a case of late Victorian madness. Did Ruskin imagine "the end of nature" a hundred years before Bill McKibben? The Swedish physicist Svante Arrhenius demonstrated the warming effect of carbon dioxide on the atmosphere in 1896, just a few years after Ruskin's scandalous London lectures on climate change.[5]

What would it be like to choose sufficiency over cornucopianism? A new, radical version of environmentalism is looking for practical solutions in the face of the Anthropocene. Bill McKibben urges us to back off from the dream of exponential growth "lightly, carefully, gently." The Transition Town network in the United Kingdom is pioneering a systematic effort to decarbonize local economies. The doyen of the Green Movement in the United States—Gus Speth—wants to reorient the American Dream toward a new measure of ecological wealth, beyond the old worldview of GDP. These radical critics warn that genuine reform requires a profound cultural transformation to counter and overcome the cornucopian beliefs that pervade contemporary economics and politics.[6]

In times like these, the Victorian notion of the stationary state serves as both an inspiration and a warning example. The first lesson is perhaps the most comforting. We have been here before. There is a long history behind the radical dreams of the present. The biographies of Ruskin, Collingwood and his family, Beever, Fleming, and Rawnsley all hint at the powerful place of the creative imagination in shaping alternatives to mass consumption. In this way they shed light on the personal experience of voluntary sufficiency, as a lived reality rather than an abstract principle. But their experiments with sufficiency also expose a number of contradictions and blind spots in the vision of a post-growth society. Ruskin's paternalism made it all too easy for him to let poor and uneducated people carry the weight of reform. He could not see why Agnes Walker and her family might want a different life than what was offered in the cottage at Lawson Park. In the hands of Hardwicke Rawnsley, the celebration of local virtue turned into a principle of vicarious sufficiency: the common people of the Lake District had to be protected from the temptations (and opportunities) of consumer society. Another fundamental contradiction concerned the place of technology. Here, the rival views of Ruskin and Rawnsley are illuminating. Can a systematic critique of industrial capitalism still embrace a measure of technological optimism? As we have seen, Ruskin rejected the use of steam engines but embraced more traditional forms of labor powered by wind and water. Rawnsley moved beyond Ruskin's apocalyptic fear of steam to imagine how trains and chimneys might serve the ideals of preservation and sufficiency. Yet too strong a faith in technological innovation easily slips into the kind of magical thinking that expects a technical fix for every environmental crisis. Our green Victorians offer no clear remedy to this problem. But surely any attempt to solve it must begin with the more fundamental question, which they knew so well: what makes a good life?

Acknowledgments

This book has been a labor of love, in more than one sense. Our project came to life long before we realized it, during a visit to the Lakes more than a decade ago. A chance encounter with the house and gardens of the Ruskin Museum at Brantwood left a lasting impression. A few years later, our much-missed friend, the late Emile Perreau-Saussine, gave us a beautiful copy of *Unto This Last* and *The Two Paths*, which planted a seed of interest in Ruskin's political economy. Reading Ruskin became a happy distraction from other, more pressing claims on our time. It was only when we connected the history of the Arts and Crafts movement to current debates about the environment and sustainable living that the argument of the book started to take shape in earnest, setting us on a course of research from Cumbria and London to New Haven and Pasadena.

We are keenly aware that we are interlopers in the world of Ruskin scholarship, conscious of our shortcomings in grappling with the formidable oeuvre of Ruskin's writings. But we have been extremely fortunate in receiving friendly encouragement from experts in the field, including Howard Hull, the director of the Brantwood museum, and Stephen Wildman, the director of the Ruskin Library and Research Center at Lancaster University. We have also benefited immensely from the superb scholarship on Lakeland preservationism and Arts and Crafts, in particular the work of Harriet Ritvo, Sara Haslam, Matthew Townend, and Jennie Brunton. Unfortunately Mark Frost's important book on the Guild of St. George reached us too late to be included in our argument.

We owe a special debt of gratitude to the following persons for their help in solving particular problems of research. In Coniston, Marguerite and Graham Aldridge showed us the grounds of the Thwaite and shared a map of Susanna Beever's garden. At Brantwood, the director Howard Hull and head gardener Sally Beamish answered many questions about the economy of the household and the history of the landscape. Vicky Slowe at the Ruskin Museum in Coniston helped us identify numerous individuals in several photographs. At the Ruskin Library, Stephen Wildman provided advance access to his article on Ruskin and the Druces. Sue Hodkins of the Huntington Library helped unravel a decades-old labeling error in the Beever correspondence. At Abbot Hall, Nick Rogers and James Arnold gave us access to the treasure of the Collingwood correspondence and the family magazine *Nothing Much*. GIS specialist Chieko Maeni of the University of Chicago assisted in making the map of Ruskin's Lake District.

We gratefully acknowledge the kind permission to quote from sources held by a number of archives and libraries in Britain and the United States, including the National Library of Scotland, the British Library, the Huntington Library, the Beinecke Library at Yale University, the Ruskin Library and Research Center at Lancaster University, the Museum of Lakeland Life and Industry at Abbot Hall in Kendal, and the Cumbria Archive Centres in Kendal and Carlisle (formerly known as the Cumbria Record Office).

One of the greatest pleasures of this project has been the opportunity to explore the visual culture of the Arts and Crafts movement. Howard Hull generously shared with us pictures of Ruskin and Brantwood. Rights to reproduce photographs of the Pepper family and the Ruskin household on the ice were granted us by kind permission of Vicky Slowe at the Ruskin Museum in Coniston. James Arnold, Teresa Smith, and David Boucher granted us permission to reproduce images from the Collingwood family archive at Abbot Hall in Kendal, including the family magazine *Nothing Much*. Teresa Smith also supplied Collingwood family photos from her private collection. Ruskin's painting of the peacock feather first came to our attention by way of Howard Hull. We were able to procure a digital copy and republication rights from the Collection of the Guild of St. George, Museums Sheffield, with the help of Louise Pullen (Ruskin curator), Denise Butler, and Julie Taylor, whose quick, comprehensive, and cheerful response was much appreciated.

Many people have encouraged our fascination with Ruskin and the Lakes over the years. Jens Spinger and Bryan Garsten accompanied Fredrik on a first trip to Coniston in the fall of 1996. At the ASEH conference in Toronto, Don Worster offered kind encouragement. In Paris, Charles-

François Mathis took us out for a sumptuous dinner and gave us splendid advice. Kristin Boyce proofread the introduction and never failed to lend a sympathetic ear. Anya Zilberstein and Emily Pawley looked over a draft of the *Nothing Much* chapter and gave us excellent feedback. Tara Zahra and Matthew Edney discussed the history of childhood and family magazines with us. Ian Desai helped us with the Gandhi and Ruskin connection. Our dear friend Mark Fiege invited us to give our first joint talk about Ruskin and Lakeland Arts and Crafts at Montana State University as guest speakers in the Wallace Stegner Chair program. Thanks to Mark, Janet Ore, Dean Nic Rae, Philip Gaines, Susan Cohen, Catherine Dunlop, Susan Kollin, Michael Reidy, Tim LeCain, Dale Martin, Brett Walker, Bob Rydell, and all the other wonderful people we met in Montana.

A number of friends and colleagues have offered moral support during the long road of research and writing drafts, including Erin Hollaway Palmer and Brian Palmer, Jacob Blakesley, Irina Ruvinsky, Jackie Feke, Anders Brunstedt, Julia Adeney Thomas, Emily Osborn, Dipesh Chakrabarty, Chris Smout, Deborah Cohen, John Brewer, Maura Capps, Paul Cheney, and Carl Wennerlind.

An earlier version of chapter 3 entitled "The Sufficient Muse" was published in *The Eighth Lamp: Ruskin Studies Today*. Vicky would like to thank the editors Anuradha Chatterjee and Laurence Roussillon-Constanty as well as the two anonymous reviewers. Fredrik presented part of the argument of chapter 1 at a conference on historicism and the human sciences organized by Mark Bevir at Berkeley.

Our editor at the University of Chicago Press, Karen Merikangas Darling, and her assistant, Evan White, have given us marvelous backing and assistance at every step of the way. We are also deeply grateful to the two anonymous reviewers of the manuscript for their exemplary feedback. Finally, Marianne Tatom succeeded in the rare feat of making the last stage of copyediting and proofreading both pleasant and easy.

This book could never have been written without the unwavering support and abiding love of our parents, Siv Jonsson and Phyllis and David Albritton. We dedicate it to them with great affection and gratitude.

Abbreviations

AH	Abbot Hall Art Gallery and Archive, Kendal
AWS	Arthur Simpson
Beinecke	The Beinecke Library, Yale University
BL	The British Library
CACC	Cumbria Archive Centre Carlisle (formerly Cumbria Record Office Carlisle)
CACK	Cumbria Archive Centre Kendal (formerly Cumbria Record Office Kendal)
CWAAS	Cumberland and Westmorland Antiquarian and Archeological Society
EMDC	Edith Mary "Dorrie" Collingwood
HL	The Huntington Library, Pasadena
JR	John Ruskin
KSIA	Keswick School of Industrial Art
NLS	The National Library of Scotland
NM	*Nothing Much*
RL	Ruskin Library, Lancaster
SB	Susanna Beever
WGC	William Gershom Collingwood
Works	John Ruskin, *The Works of John Ruskin*, eds. E. T. Cook and Alexander Wedderburn (London: George Allen, 1903-1912)

Notes

INTRODUCTION

1. Albert Fleming, "Hand Spinning and Weaving in Westmoreland," *Century Illustrated Monthly Magazine*, no. 1 (November 1889), 522–23.
2. Ibid.
3. See Richard D. Altick, "Nineteenth-Century English Best-Sellers: A Third List," in *Studies in Bibliography*, vol. 39 (1986), 235–41; Aileen Fyfe, *Steam-Powered Knowledge: William Chambers and the Business of Publishing 1820–1860* (Chicago: University of Chicago Press, 2012), xvi, 9, 55–64. *Times*, typecasting, Wick's, Linotype: Michael Twyman, *The British Library Guide to Printing History and Techniques* (London: British Library, 1998), 70, 73, 75.
4. Seth Koven, *Slumming: Sexual and Social Politics in Victorian London* (Princeton, NJ: Princeton University Press, 2014); ibid., *The Match Girl and the Heiress* (Princeton, NJ: Princeton University Press, 2014); W. T. Stead, "The Maiden Tribute of Modern Babylon," *Pall Mall Gazette*, July 1885; Wanda F. Neff, *Victorian Working Women: An Historical and Literary Study of Women in British Industries and Professions 1832–50* (Abingdon: Routledge, 2006), 38.
5. H. H. Warner, *Songs of the Spindle & Legends of the Loom* (London: N. J. Powell & Co., 1889), 7. All photographs of the exterior and contents of *Songs of the Spindle & Legends of the Loom* were taken using a private copy owned by the authors.
6. Fleming's move to the Lakes, Sara E. Haslam, *John Ruskin and the Lakeland Arts Revival, 1880–1920* (Cardiff: Merton Priory Press, 2004), 17; Ruskin's resting place, see Tim Hilton, *John Ruskin* (New Haven, CT: Yale University Press, 2002), 874.
7. Warner, *Songs of the Spindle*, 7–8. Bill McKibben, *Eaarth: Making a Life on a Tough New Planet* (New York: Henry Holt, 2010), 102–3; Clive Hamilton, *Affluenza: When Too Much Is Never Enough* (Crows Nest, Australia: Allen & Unwin, 2005); Diane Dumanoski, *The End of the Long Summer: Why We Must Remake Our Civilization to Survive on a Volatile Earth* (New York: Broadway Books, 2010); Tim Jackson, *Prosperity without Growth: Economics for a Finite Planet* (New York: Routledge, 2011); Juliet Schor, *Plenitude: The New Economics of True Wealth* (New York: Penguin, 2010).
8. We note in passing that sufficiency as a principle can be integrated into a wide range of different cultures. From this perspective, it may be more appropriate to speak in plural of *cultures of sufficiency*.
9. Thomas Princen, *The Logic of Sufficiency* (Boston: MIT Press, 2005), 2; Schor, *Plenitude*, 6; Hamilton, *Affluenza*, 7; Robert Skidelsky and Edward Skidelsky, *How Much Is Enough: Money and the Good Life* (New York: Other Press, 2012), 3. Cf. David E. Shi, *The Simple Life: Plain Living and*

High Thinking in American Culture (Athens: University of Georgia Press, 2007); David A. Crocker and Toby Linden, eds., *Ethics of Consumption: The Good Life, Justice, and Global Stewardship* (Lanham, MD: Rowman & Littlefield, 1998); Richard Easterlin, "The Economics of Happiness," *Daedalus*, vol. 133, no. 2 (Spring 2004), 26-33.

10. Ruskin did not use the specific word *sufficiency* in his political economy and aesthetic writings, but he was keenly aware of the concept and expressed it with words like *contentment*; see, for example, *Modern Painters, in* John Ruskin, *The Works of John Ruskin*, eds. E. T. Cook and Alexander Wedderburn (London: George Allen, 1903-1912), 7.425-26: "So that the things to be desired for man in a healthy state, are that he should not see dreams, but realities; that he should not destroy life, but save it; and that he should be not rich, but content." For further analysis of this passage, see chapter 1.

11. A notable exception to the neglect of the theme of the simple life is the scholarship on Henry David Thoreau; for example, see Donald Worster, *Nature's Economy (Cambridge, UK: Cambridge University Press, 1994)*; Lawrence Buell, *The Environmental Imagination: Thoreau, Nature Writing, and the Formation of American Culture* (Cambridge, MA: Harvard Belknap, 1996); David Robinson, *The Natural Life: Thoreau's Worldly Transcendentalism* (Ithaca, NY: Cornell University Press, 2004). Romantic sensibility and natural history: see Richard Grove, *Green Imperialism; Colonial Expansion, Tropical Island Edens and the Origins of Environmentalism, 1600-1860* (Cambridge, UK: Cambridge University Press, 1995); Worster, *Nature's Economy*; idem, *A Passion for Nature: The Life of John Muir* (Oxford: Oxford University Press, 2011); T. C. Smout, *Nature Contested; Environmental History in Scotland and Northern England since 1600* (Edinburgh: Edinburgh University Press, 2000); Aaron Sachs, *The Humboldt Current: Nineteenth-Century Exploration and the Roots of American Environmentalism* (New York: Penguin, 2007); Laura Dassow Walls, *The Passage to Cosmos: Alexander von Humboldt and the Shaping of America* (Chicago: University of Chicago Press, 2011); William Wordsworth, *The Prelude: the Four Texts (1798, 1799, 1805, 1850)*, ed. Jonathan Wordsworth (London: Penguin, 1995); Karl Jacoby, *Crimes against Nature: Squatters, Poachers, Thieves, and the Hidden History of American Conservation* (Berkeley: University of California Press, 2003).

12. Conrad Totman, *The Green Archipelago: Forestry in Pre-Industrial Japan* (Athens: Ohio University Press, 1998); Karl Appuhn, *A Forest on the Sea: Environmental Expertise in Renaissance Venice* (Baltimore, MD: Johns Hopkins University Press, 2009); Grove, *Green Imperialism*; Gregory Barton, *Empire Forestry and the Origins of Environmentalism* (Cambridge, UK: Cambridge University Press, 2007); Fredrik Albritton Jonsson, *Enlightenment's Frontier: The Scottish Highlands and the Origins of Environmentalism* (New Haven, CT: Yale University Press, 2013); Elinor Ostrom, *Governing the Commons: The Evolution for Institutions of Collective Action* (Cambridge, UK: Cambridge University Press, 1990); J. M. Neeson, *Commoners: Common Right, Enclosure and Social Change in England, 1700-1820* (Cambridge, UK: Cambridge University Press, 1996); Paul Warde, *Ecology, Economy, and State Formation in Early Modern Germany* (Cambridge, UK: Cambridge University Press, 2006); Brian Donahue, *The Great Meadow: Farmers and the Land in Colonial Concord* (New Haven, CT: Yale University Press, 2007); Tine de Moor, "The Silent Revolution: A New Perspective on the Emergence of Commons, Guilds, and Other Forms of Collective Corporate Action in Western Europe," *IRSH*, vol. 53 (2008), Supplement, 179-212.

13. Björn-Ola Linnér, *Return of Malthus: Environmentalism and Postwar Population-Resource Crises* (Isle of Harris: White Horse Press, 2004); Thomas Robertson, *The Malthusian Moment: Global Population Growth and the Birth of American Environmentalism* (New Brunswick, NJ: Rutgers University Press, 2012), 36-37; Paul Sabin, *The Bet: Paul Erlich, Julian Simon, and Our Gamble over Earth's Future* (New Haven, CT: Yale University Press, 2013); Jacob Darwin Hamblin, *Arming Mother Nature: The Birth of Catastrophic Environmentalism* (Oxford: Oxford University Press, 2013); Libby Robin, Sverker Sörlin, and Paul Warde, *The Future of Nature: Documents of Global Change* (New Haven, CT: Yale University Press, 2013); E. F. Schumacher, *Small Is Beautiful: Economics as if People Mattered* (New York: Harper & Row, 1973), 33.

14. William Nordhaus, *The Climate Casino: Risk, Uncertainty, and Economics for a Warming World* (New Haven, CT: Yale University Press, 2013); Clive Hamilton, *Earthmasters: The Dawn of the Age of Climate Engineering* (New Haven, CT: Yale University Press, 2013). In addition to nn. 7-8, cf. Steven Gardiner, *A Perfect Moral Storm: The Ethical Tragedy of Climate Change* (Oxford: Oxford University Press, 2013); Paul Gilding, *The Great Disruption: Why the Climate Crisis Will Bring on the End of Shopping and the Birth of a New World* (London: Bloomsbury, 2011); Gus Speth, *The Bridge at the End of the World: Capitalism, the Environment and Crossing from Crisis to Sustainability* (New Haven, CT: Yale University Press, 2009).

15. *Works*, 29:249 (October 1877).

16. Jonathan Bate, *Romantic Ecology: Wordsworth and the Environmental Tradition* (New York: Routledge, reprint, 2011); Michael Wheeler, ed., *Ruskin and the Environment* (Manchester, UK: Manchester University Press, 1995); James Winter, *Secure from Rash Assault: Sustaining the Victorian Environment* (Berkeley: University of California Press, 2002); Harriet Ritvo, *The Dawn of Green: Manchester, Thirlmere and Modern Environmentalism* (Chicago: University of Chicago, 2009); Peter Harman, *The Culture of Nature in Britain, 1680-1860* (New Haven, CT: Yale University Press, 2009); Karen Welberry, "H. D. Rawnsley, W. G. Collingwood, and the German Miners of Keswick 1565-1645: An Early Quarrel over Conservationist Paradigms in Lakeland," *Collingwood and British Idealism Studies*, vol. 10, 1-22; Charles-François Mathis, *In Nature We Trust: Les paysages anglais à l'ère industrielle* (Paris: Presses de l'université Paris-Sorbonne, 2010).

17. *Unto This Last* was published in the form of newspaper articles in 1860 and as a book in 1862. *Unto This Last: Four Essays on the Principles of Political Economy* (London, 1862); see Ruskin, *Works*, 17: *Unto This Last, Munera Pulveris, Time and Tide* (London, 1905). *Fors Clavigera* was published in installments 1871-84 and then gathered together into book form; see Ruskin, *Works*, 27-29: *Fors Clavigera: Letters to the Workmen and Labourers of Great Britain* (London, 1907). See also Dinah Birch, ed., *Fors Clavigera* (Edinburgh: Edinburgh University Press, 2000); ibid., ed., *Ruskin and the Dawn of the Modern* (Oxford: Oxford University Press, 1999).

18. Donald Worster, *A Passion for Nature*; Tim Hilton, *John Ruskin*.

19. Slave ship: Ruskin, *Works*, 3.571-73; George P. Landow, *Images of Crisis: Literary Iconology, 1750 to the Present* (Boston: Routledge, 1982), 196.

20. John Muir, *San Francisco Daily Evening Bulletin*, November 18, 1875.

21. On the Jura: Ruskin, *Works*, 8.222-24; W. G. Collingwood, *The Lake Counties* (London: J. M. Dent and Sons, 1902), 4. Compare *Works*, 17.111: "The desire of the heart is also the light of the eyes. No scene is continually and untiringly loved, but one rich by joyful human labour; smooth in field; fair in garden; full in orchard; trim, sweet, and frequent in homestead; ringing with voices of vivid existence." *Works*, 20.36: "[N]o race of men which is entirely bred in wild country, far away from cities, ever enjoys landscape." On Muir's wilderness, see Donald Worster, *A Passion for Nature*, 186-88, 211, 215. For an excellent cultural and intellectual history of mountains and mountain climbing, see Peter H. Hansen, *Summits of Modern Man: Mountaineering after the Enlightenment* (Cambridge, MA: Harvard University Press, 2013).

22. We borrow the term *ethics of consumption* from John M. Craig, *John Ruskin and the Ethics of Consumption* (Charlottesville: University of Virginia Press, 2006). On Ruskin's anthropocentric views, see Sachs, *The Humboldt Current*, 193. Sachs goes on to note that Ruskin "could also appreciate the geological sublime," but he does not spell out the marked level of fear and even aversion (as opposed to wonder) that colored Ruskin's perception of uninhabited regions. This fear is palpable in his description of the Alps in particular at *Works* 6.295-96. For Michael Wheeler, Ruskin's Christian anthropocentrism sets him apart from the biocentric perspective of modern ecological thought; see Wheeler, *Ruskin's God* (Cambridge, UK: Cambridge University Press, 1999), 278. Note also Wheeler's discussion of how Ruskin grew "bored with the High Alps" and embraced the life of the city of Turin in the summer of 1858; ibid., 125.

23. Linda Lear, *Beatrix Potter: A Life in Nature* (New York: St. Martin's Griffin, 2008), 72.

24. Compare with Ritvo's use of *preservation* in *The Dawn of Green*.

CHAPTER ONE

1. Peter Baxter to Joan Severn (August 30, 1887, and September 1, 3, and 4, 1887), RL.
2. John Ruskin, "Cheque to Kate Raven," April 28, 1887, RL.
3. Homespun dress: James Dearden, *John Ruskin: A Life in Pictures* (Sheffield: Sheffield Academic Press, 1999), 14, 18. Satisfaction, *Works*, 7.426; cf. *Works*, 16.18, "[M]an's labor, well applied, is always amply sufficient to provide him during his life with all things needful to him." On the retail revolution and Maple's, see Deborah Cohen, *Household Gods* (New Haven, CT: Yale University Press, 2006), 44, 51; Caligula, *Works*, 4.61–62.
4. *Works*, 17.84.
5. See the twenty-ninth letter of the *Fors Clavigera* from 1873, *Works*, 27.527. William Stanley Jevons, *The Coal Question: An Inquiry Concerning the Progress of the Nation, and the Probable Exhaustion of Our Coal Mines* (London: Macmillan, 1865).
6. *Works*, 2.265 (and n. 1). See also *The Poems of John Ruskin, Now First Collected from Original Manuscript and Printed Sources*, ed. W. G. Collingwood (Sunnyside: George Allen, 1891).
7. Tim Hilton, *John Ruskin* (New Haven, CT: Yale University Press, 2002): Bible, 13, 19; Wordsworth, 73; reputation, 107.
8. For a survey of social policy in the period, see Philip Harling, *The Modern British State: An Historical Introduction* (Oxford: Blackwell, 2001), chapter 3.
9. *Works*, 27.528–29.
10. *Works*, 27.529–30. For the wider conflict over city life, see Tristram Hunt, *Building Jerusalem: the Rise and Fall of the Victorian City* (London: Phoenix, 2005[2004]).
11. *Works*: monkeys, 27.531; operative, 10.94–95; Adam Smith, *The Wealth of Nations*, 2 vols. (Oxford: Oxford University Press, 1976), vol. 2, 782.
12. *Works*, 17.44.
13. Tim Hilton, *John Ruskin*, 493–95. For back-to-the-land Victorians, see Charles-François Mathis, *In Nature We Trust: Les paysages anglais à l'ère industrielle* (Paris: Presses de l'université Paris-Sorbonne, 2010).
14. *The Economist of Xenophon*, ed. John Ruskin, trans. Alexander Wedderburn and W. G. Collingwood (Kent: George Allen, 1876), 49–50 (*Works*, 31.58); see also Willie Henderson, "Xenophon, Ruskin, and Economic Management," in *John Ruskin's Political Economy* (New York: Routledge, 2000), 64–85.
15. John Stuart Mill, *Principles of Political Economy*, ed. Sir William Ashley (New York: Augustus M. Kelley, 1987[1848]), 751. Cf. Graham Macdonald, "The Politics of the Golden River: Ruskin on Environment and the Stationary State," *Environment and History*, vol. 18 (2012), 125–50.
16. *Works*, 17.113.
17. *Works*: value of a thing independent, 17.85; intrinsic value, sheaf of wheat, 17.153; romantic, 17.94.
18. *Works*: infinite wealth, 28.14–16; beer and pipes, 29.18.
19. *Queen of the Air: Being a Study of the Greek Myths of Cloud and Storm* (New York: John Wiley,1871), 130; see also *Works*, 19.398; Jevons, *The Coal Question*. James Clark Sherburne reads Ruskin as a prophet of social abundance but misses the crucial significance of "satisfaction" and "contentment"; see *John Ruskin or the Ambiguities of Abundance: A Study in Social and Economic Criticism* (Cambridge, MA: Harvard University Press, 1972), 73. Contrast with P. D. Anthony, *John Ruskin's Labour: A Study of Ruskin's Social Theory* (Cambridge, UK: Cambridge University Press, 1983), 94.
20. No wealth but life, *Works*, 17.105; posterity, H. D. Rawnsley, *Ruskin and the English Lakes* (Glasgow: James MacLehose, 1902), 211; consumption, *Works*, 17.98; greatest number, *Works*, 17.105; simpler pleasures, *Works*, 17.112; not socialist, *Works*, 17.106–7, note.
21. See Hilton, *John Ruskin*, 766, for how "some papers ridiculed Ruskin's views."
22. *Works*: plague-wind, 34.31; malignant, 34.34; N. England to Sicily, 34.32; intermittent,

34.34-35; always dirty, 34.51. See also Peter Thorsheim, *Inventing Pollution: Coal, Smoke, and Culture in Britain since 1800* (Athens: Ohio University Press, 2006), 55.

23. *Works*: like aspens, 34.34; trembling continuously, 34.31.

24. *Works*: smoke-cloud, Manchester devil, 34.36-37; dead men's souls, 34.33; blanched sun, 34.40. It is worth stressing here that Ruskin slipped in and out of the role of a world-weary prophet. However much the Storm Cloud weighed on his mind, he still took great pleasure in his trips to London, visiting museums, shops, and having tea with friends. Ruskin's "prophetic function," W. G. Collingwood observes, "was not all sackcloth and ashes." See W. G. Collingwood, *The Life of John Ruskin* (Boston: Houghton Mifflin Co., 1893), 377, note.

25. *Works*: Blasphemed, 34.40; blasphemy defined, 34.72; injustice, 34.40; Solomon, usury, and swindling, 34.74; cheerfulness, 34.41. Cf. Michael Wheeler, *Ruskin's God*, 273; Dinah Birch, "Fallen Nature: Ruskin's Political Apocalypse," in John Ginall and H. Gustav Klaus, eds., in association with Valentine Cunningham, *Ecology and the Literature of the British Left: The Red and the Green* (Farnham, UK: Ashgate, 2012). See also Thorsheim, *Inventing Pollution*, 56-57. Although Britain and Europe were at the center of the Storm Cloud, Ruskin alluded several times to the global effects of the crisis. *Works*: half around the world, 34.39; all foreign nations, 39.40; the Empire of England, 39.41; the Kosmos and the Earth, 39.42.

26. *Works*, 34.31.

27. *Works*: dunghills, 7.377; axe, 7.387; charnel house, 7.387; cf. also Raymond Fitch, *The Poison Sky: Myth and Apocalypse in Ruskin* (Athens: Ohio State University Press, 1982), 391.

28. Vitiate, *Works*, 27.91; Mill, *Principles of Political Economy*, 750; Donald Winch, *Wealth and Life: Essays on the Intellectual History of Political Economy in Britain, 1848-1914* (Cambridge, UK: Cambridge University Press, 2009), 96.

29. *Works*: extremity, 36.454; Bionnassay, 37.145; violent, 18.357; umbered, 19.293.

30. John Tyndall, *Forms of Water: In Clouds and Rivers, Ice and Glaciers* (New York, 1872), xi; *Works*, one third, 27.635-36; blasphemous science, 34.73; ironically, Ruskin's arch-enemy, John Tyndall, was the first scientist to verify experimentally the role of carbon dioxide as a greenhouse gas. For a historically sensitive account of Tyndall, see Joshua P. Howe, "Getting Past the Greenhouse: Taking Nineteenth-Century Climate Science on Its Own Terms," in Bernard Lightman and Michael Reidy, eds., *The Age of Scientific Naturalism: John Tyndall and His Contemporaries* (London: Pickering and Chatto, 2014). In an 1843 letter, Ruskin argued that the luxuriant vegetation of prehistoric times had been the result of very high levels of carbonic acid (carbon dioxide) in the atmosphere but that this "carbon" had since been "deposited" in the earth to form the great "coal-fields" of the world, see *Works*, 1.487. Cf. Robert Chambers, *Vestiges of the Natural History of Creation*, ed. James Secord (Chicago: University of Chicago Press, 1994), 57.

31. *Works*: deterioration, 28.488.

32. *Works*: deforestation, 29.332; increased rainfall, 29.348. Eugene Viollet-le-Duc, *Mont Blanc* (London: Sampson Low, Marston, Searle, and Rivington, 1877), 341-42. Ruskin's concept of climate change was indebted to a long-standing discourse on deforestation; see George Perkins Marsh, *Man and Nature: or Physical Geography as Modified by Human Action* (New York, 1864), 214-16; Richard Grove stresses the colonial origins of this fear in *Green Imperialism: Colonial Expansion, Tropical Island Edens and the Origins of Environmentalism, 1600-1860* (Cambridge, UK: Cambridge University Press, 1995); Jean-Baptiste Fressoz and Fabien Locher, "Modernity's Frail Climate: A Climate History of Environmental Reflexivity," *Critical Inquiry*, vol. 38, no. 3 (Spring 2012), 579-98; Fredrik Albritton Jonsson, "Climate Change and the Retreat of the Atlantic: The Cameralist Context of Pehr Kalm's Voyage to North America 1748-51," *William and Mary Quarterly*, vol. 72, no. 1 (January 2015), 99-126. For the development of climate science in recent times, see Paul N. Edwards, *A Vast Machine: Computer Models, Climate Data, and the Politics of Global Warming* (Cambridge, MA: MIT Press, 2010).

33. All snows wasted, *Works*, 37.408; W. G. Collingwood, *Ruskin Relics* (New York: T. Y. Crowell & Co., 1904), 55-59; climate of Europe growing damper, *Works*, 34.675-76; Hilton, *John Ruskin*, 728.

34. *Works*, 33.404.

35. For the Little Ice Age, see Geoffrey Parker, *Global Crisis: War, Climate Change and Catastrophe in the Seventeenth Century* (New Haven, CT: Yale University Press, 2013); on alpine glaciers, see Jean M. Grove, *The Little Ice Age* (New York: Routledge, 2008), chapters 6 and 11. For CO_2 and natural variability, see Will Steffen, Paul J. Crutzen, and John McNeil, "The Anthropocene: Are Humans Now Overwhelming the Great Forces of Nature?" *AMBIO: A Journal of the Human Environment*, vol. 36, no. 8 (2007), 616.

36. Helen Gill Viljoen, *The Brantwood Diary of John Ruskin* (New Haven, CT: Yale University Press, 1971), 146, 58.

37. John Illingworth, "Ruskin and Gardening," *Garden History*, vol. 22, no. 2, "The Picturesque" (Winter 1994), 218–33; David Ingram, *The Gardens at Brantwood: Evolution of John Ruskin's Lakeland Paradise* (London: Pallas Athene & the Ruskin Foundation, 2014).

38. *Works*: wood-woman, 37.554; wood-woman came every day, 35.xxvi; on fuel prices, see for instance 27.527; smoke a purer blue, 29.273.

39. *Works*, wood versus coal, 28.470.

40. Illingworth, "Ruskin and Gardening," 231. Wheat, cranberries, Ingram, *The Gardens at Brantwood*, 68.

41. Ingram, *The Gardens at Brantwood*, 26–30, 45.

42. For Ruskin's views regarding the state, see Jose Harris, "Ruskin and Social Reform," in Dinah Birch, *Ruskin and the Dawn of the Modern*, 7–33; Stuart Eagles, *After Ruskin: The Social and Political Legacies of a Victorian Prophet, 1870-1920* (Oxford: Oxford University Press, 2011), chapter 1, passim; Michael Wheeler, *Ruskin's God*, chapter 7, passim.

43. Hilton, *John Ruskin*: proposal, 380–83; courtship and breakdown, 459; Rose's death, 580–86.

CHAPTER TWO

1. Glass beads, *Works*, 10.197.

2. Noble, *Works*, 10.196–97; Greek vs. Gothic, 10.202.

3. James S. Dearden, *John Ruskin's Guild of St. George* (Bembridge: Guild of St. George, 2010), 7–8.

4. License, museum, Dearden, *Guild of St. George*, 8–10. Wind and water, *Works*, 30.48.

5. Sue King, *A Weaver's Tale: The Life & Times of the Laxey Woollen Industry 1860-1010* (Laxey: St. George's Woollen Mills Ltd., 2010): Egbert Rydings, 28–30; £25 check, no hand spinning, 32; purest quality, 43.

6. King, *A Weaver's Tale*: pure wool, 43; wool washed, natural soap, 45; metal spikes, Card Room Fog, mechanized spinning mule, water-powered looms, 46.

7. King, *A Weaver's Tale*: liquid slime, 45; cancer, rashes, 48; other health problems, 49.

8. King, *A Weaver's Tale*: £200, two decades, 55; end of guild involvement, 73; twenty-first century, 149 (cp. 120ff).

9. See Sara Haslam, *John Ruskin and the Lakeland Arts Revival, 1880-1920* (Cardiff: Merton, 2004). Haslam sets out the various arts and crafts industries flourishing elsewhere at the time on pp. 1–2. She also sharply distinguishes between them and the unique effect that Ruskin's "close presence" had on the Lakeland industries, as well as the influence of the Cumberland landscape itself. Jennie Brunton, *The Arts & Crafts Movement in the Lake District: A Social History* (Lancaster: University of Lancaster, 2001). NB: Both Haslam and Brunton are authorities on many figures and issues discussed in this chapter. We cite frequently from both but do not always cross-cite both works at once.

10. Haslam, in *Lakeland Arts*, 32, makes this point, distinguishing between Morris & Co. as an arts and crafts industry and the "true Ruskinian industry" that promotes working from home.

11. Haslam, *Lakeland Arts*, 87-89. H. D. Rawnsley, *Ruskin and the English Lakes* (Glasgow: James MacLehose and Sons, 1902), 115-18. Brunton, *Arts & Crafts*, 91-92.

12. Haslam, *Lakeland Arts*: Lake Artists' Society, 8-9; Simpson in 1885, 138; wood-carving classes and exhibitions, 138-40. Annie Garnett, 150-68 (influence of nature 151, 157-60). Brunton, *Arts & Crafts*: Simpson's death, 51; World War Two, 52.

13. Haslam, *Lakeland Arts*, 12.

14. Stephen Wildman and Sheila Clark, "Ruskin and the Druces: A Visit to Brantwood in 1884," *Friendly Neighbor: Brantwood Newsletter* (2011). The date of the entry is Brantwood Sept 25th 1884, or #30 from the transcription of the diary entries of Walter Druce. Special thanks to Stephen Wildman for allowing us to see an advance copy.

15. *Works*, 30.328.

16. Hates work, JR to Fleming, April 26, 1885 (quoting Fleming), Yale Beinecke Library. On Fleming's correspondence with JR, Haslam, *Lakeland Arts*, 15-16. Albert Fleming, *In the House of Rimmon* (London: Partridge & Cooper, 1873), 23, 38.

17. Cf. Albert Fleming, "Albrecht Dürer," *The English Illustrated Magazine* (London: Macmillan and Co., 1890), 567-87. JR lent Dürer, JR to Fleming, May 23, 1886, Yale Beinecke Library. Haslam, *Lakeland Arts*: railways, pollution, 37-38, where Haslam cites *Works*, 26.121-22, on Ruskin's view of the pollution of Derwent and Coniston Water. The phrase *rash assault* comes from William Wordsworth's sonnet "On the Projected Kendal and Windermere Railway"; see also James Winter, *Secure from Rash Assault: Sustaining the Victorian Environment* (Chicago: University of Chicago Press, 2002).

18. Haslam, *Lakeland Arts*: spinning tradition, 4; lace, feminine plough, 27-28. See also *Works*: feminine plough, 16.395; lace, 16.157-58. Blown together, homemade, Fleming, *In The House of Rimmon*, 6-7.

19. Marian Twelves learned to spin first, Haslam, *Lakeland Arts*, 19 (but note that Fleming omits Twelves's role here from his articles, saying he learned himself from an elderly lady). Albert Fleming, "Revival of Hand Spinning and Weaving in Westmoreland," *Century Illustrated Monthly Magazine*, vol. 37 (1889), 521-27: bought back thread, 525; Giotto, 526,

20. Fleming, "Revival of Hand Spinning and Weaving in Westmorland," 527.

21. In addition to the articles already mentioned, an early published account: Albert Fleming, "The Spinning Wheel," *The Standard*, October 30, 1883, p. 5. Popularize, Fleming to W. H. Hills, March 11, 1884 (DSO 24/7/2), CACC.

22. Bleaching, what factory, *Works*, 28.66. See also Haslam, *Lakeland Arts*, 18-19, but the italicized comment beginning *"what factory . . ."* appears to be the correspondent's, not Ruskin's. Lichens, natural dyes, Brunton, *Arts & Crafts*, 76, 79; also Haslam, *Lakeland Arts*, 34. Regionalism, nature-inspired, Haslam, *Lakeland Arts*, 31-33.

23. Haslam, *Lakeland Arts*: items sold, 30; Laxey, 42, where Haslam links Laxey's waterpowered mill to their lack of decorated items—that is, to their lack of emphasis on handcrafted, individually designed goods; decorations, 31; napkins, etc., 32; natural imagery, invalid, 33; an old lady, 34; Ruskin Lace, 41; pincushions, 43.

24. Fleming, *The House of Rimmon*, 30, 31. The following stories authored by "Albert Fleming" in the 1880s and 1890s are most certainly by the Albert Fleming described in this chapter since they contain references to Westmoreland or Gray's Inn, where Fleming worked: "A Westmorland Story," *Gentleman's Magazine* (December 1897), 521-36 (regarding unwed mothers); "A Romance of Gray's Inn: A Story," *Gentleman's Magazine*, vol. 54 (1894), 217ff; "Sally," *Gentleman's Magazine* (March 1891), 217ff; "The Importunate Widow," *Cornhill Magazine*, vol. 55 (May 1887), 515ff (which tells the story of a temptress who gains dangerous knowledge of legal matters. The ensuing quandary is set right by a quick-witted daughter who makes no dangerous attempts at self-education); and "A Literary Venture" (sometimes miscataloged as "A Literary Centure"), *Belgravia*, vol. 68, no. 271 (May 1889), 312-23. In this last story, a woman decides she is going to write fiction herself. Yet, instead of inventing characters, she foolishly makes use of barely dis-

guised versions of all her friends and family. The story causes an uproar and brings shame to her husband. Though George Eliot is referred to in the story as a remarkable phenomenon—a woman who in fact earns more than most men—Fleming suggests she is merely an exception to the rule. The main character, Mrs. Lovell, is by contrast "the ordinary healthy type of English matron, quite ignorant of art and literature, but entirely satisfied with herself, her children, and husband" (313). Fleming's story also seems to criticize what he perceives as women's general tendency to confuse money with genuine value. For Fleming's positive views on women (especially against Ruskin), see Haslam, *Lakeland Arts*, 61–62. Kind friend, Brunton, *Arts & Crafts*, 62.

25. Profits divided, *Works*, 30.330. Brunton, *Arts & Crafts*, 69–72, where the quarrel is covered in detail; economic principles, more independence, 99. Haslam, *Lakeland Arts*, 51–52, also covers the quarrel, noting that it sprang up from "some divergence from Ruskin's teachings."

26. Twelves read Ruskin, JR to Twelves, January 16, 1886 (T127/c4), RL; shortcomings, Haslam, *Lakeland Arts*, 42–43. Individual effort, quantity, Brunton, *Arts & Crafts*, 70, n. 46.

27. Brunton, *Arts & Crafts*: individual effort and control, 102–3; volunteers, 104, where Brunton describes them as "amateur volunteers." Princess Louise, Haslam, *Lakeland Arts*, 75.

28. Brunton, *Arts & Crafts*: silkworms, 73; Twelves leaves KSIA, 103. KSIA linen industry ended, Haslam, *Lakeland Arts*, 77.

29. See Brunton, *Arts & Crafts*, 71, where she notes that it "seems ironic that in this same year [when Marian Twelves left the LLI] particular mention should be made of personal attribution in the production of this book." It seems likely that Warner and Fleming were aware of Twelves's complaints and took them seriously. H. H. Warner, ed., *Songs of the Spindle & Legends of the Loom* (London: N. J. Powell & Co., 1889), 8–9; British Museum, 13. Heskett possibly taught Twelves: private communication from Vicky Slowe at the Ruskin Museum in Coniston.

30. Warner, *Songs of the Spindle*, 7–9. Cf. Peter Stansky, *Redesigning the World: William Morris, the 1880s, and the Arts and Crafts* (Princeton, NJ: Princeton University Press, 1985); also Fiona MacCarthy, *William Morris: A Life for Our Time* (New York: Alfred Knopf, 1995), especially 608–22. MacCarthy notes that in 1891 Morris set up the Kelmscott Press in order to create entirely handmade books, binding them with vellum (611), using ink "without chemical additives" (612), always with the "aim of beauty" (609). Ruskin's "On the Nature of the Gothic" was one of the works he reprinted (617).

31. Anonymous reviewer, *The Magazine of Art* (London: Cassell & Company, September 1889), xlviii. Patrick McDonald, "'Swift visions of centuries': Langdale Linen, *Songs of the Spindle*, and the Revolutionary Potential of the Book," *The Eighth Lamp: Ruskin Studies Today*, no. 9 (2014), 46–61 (esp. 58–59): auto-gravure, 58; watermarks, 59. The authors base their physical description of the book's paper on their own private copy, but these observations were guided by McDonald's careful examination of a different yet presumably quite similar copy. Both bear Dutch watermarks and the name Van Gelder (or at least Van-lder). Haslam, *Lakeland Arts*: Severn book cover, Pl. 23; Chicago exposition, 52; flax, 25–26. Brunton, *Arts and Crafts*: fourteen awards, 74; photo of E. Pepper, make new friends (caption), 77.

32. Fleming and Ruskin: Tim Hilton, *John Ruskin* (New Haven, CT: Yale University Press, 2002), 819–22; also John Dixon Hunt, *The Wider Sea* (London: Phoenix Giant, 1982), 398–400.

33. Burial, Hilton, 874. Pall: Haslam, *Lakeland Arts*, 82. Brunton, *Arts & Crafts*: longevity, 80; names, 104 (where she cites F. A. Benjamin, *The Ruskin Linen Industry of Keswick* [Michael Moon, 1974]).

CHAPTER THREE

1. See Tim Hilton, *John Ruskin (New Haven, CT: Yale University Press, 2002)* for: Chesneau, 772; Norton, 771; Darwin, 685. The garden description is based mainly on William Tuckwell, *Tongues in Trees and a Sermon in Stone* (London: George Allen, 1891), 109–10. Cf. apple-perfumed,

John Ruskin, *Hortus Inclusus*, ed. Albert Fleming (New York: John Wiley & Sons, 1887). Though Beever mentions both damask plum trees and gooseberry bushes in her letters, their position in the garden is surmised from the last known diagram of the garden (made in the mid-twentieth century), before it was altered to its current arrangement. The diagram was supplied by Graham and Margaret Aldridge, former owners of Thwaite Cottage and the grounds that used to belong to the Thwaite manor house. Moved to Coniston in 1827: see obituary of Miss Susanna Beever, HL; lived in Thwaite House, 1827, *A Biographical Index of British and Irish Botanists*, eds. James Britten, G. S. Boulger (London: West, Newman & Co., 1899), 196; but cf. Frederick Sessions, *Literary Celebrities of the English Lake-District* (London: Elliot Stock, 1907), 190, where he says the Beevers moved to the Thwaite House in 1831. See also "Thwaite House to Be Let," *Lancaster Gazette*, June 21, 1828, where the house and grounds are described in detail.

2. John Hayman, "John Ruskin's *Hortus Inclusus*: The Manuscript Sources and Publication History," *Huntington Library Quarterly*, vol. 52, no. 3 (1989), 363–87 (quotations on 365–67). For "plague wind," see chapter 1.

3. 900 letters: Hilton, *John Ruskin*, 553, and cf. Hayman, 385, n. 20. Seth Koven, "How the Victorians Read *Sesame and Lilies*," in John Ruskin, *Sesame and Lilies*, ed. Deborah Epstein Nord (New Haven, CT: Yale University Press, 2002), 165–204 (168); *Works*: queenly office, 18.136; "Ruskin School," 37.411 (cf. Ruskin, *Hortus Inclusus*, 82).

4. In addition to *Hortus Inclusus* and Ruskin, *Works*, sources for S. Beever's life are Hayman, "John Ruskin's *Hortus Inclusus*"; obituary of Susanna Beever (clipping from *Westmoreland Gazette*, ca. 1893), HM 62832, Huntington Library; W. G. Collingwood, preface to John Beever's *Practical Fly-Fishing, Founded on Nature* (London: Methuen and Co., 1893[1849]); and *Oxford Dictionary of National Biography*, "Women Artists in Ruskin's Circle."

5. Friar's crag, *Works*, 5.365, and cf. John Dixon Hunt, *The Wider Sea: A Life of John Ruskin* (London: Phoenix, 1998[1982]), 44. Ruskin's early education: Hilton, *John Ruskin*, 13–17.

6. Collingwood writes that "Baxter in his *British Flowering Plants*" thanks S. Beever for a "curious" specimen of pearlwort along with an "accompanying drawing" and other information she and Mary sent him (J. Beever, *Practical Fly Fishing*, xii). Collingwood notes as well on p. xiii that they are cited in J. G. Baker, *Flora of the English Lake District* (1885). Further, "M. Beever" is cited numerous times in George Luxford, *The Phytologist: A Popular Botanical Miscellany* (London: John Van Voorst, 1844); "Miss S. Beever" is cited in Thomas Moore, *A Popular History of the British Ferns* (London: Routledge, Warne, and Routledge, 1862); and "Miss M. Beever" (also "Miss Beever") is cited in Thomas Moore, *Nature-Printed British Ferns* (London: Bradbury & Evans, 1863).

7. Sometimes called *Lastraea filix-mas*. See Marilyn Ogilvie and Joy Harvey, eds., *The Biographical Dictionary of Women in Science: Pioneering Lives from Ancient Times to the Mid-20th Century* (New York: Routledge, 2000), 104.

8. See W. G. Collingwood's preface to John Beever's *Practical Fly-Fishing*, xix.

9. See H. M. Bowdler, *Fragments, in Prose and Verse: by Miss Elizabeth Smith, Lately Deceased* (London: Cadell & Davies, 1809). Smith learned nine languages, including Greek, Latin, Arabic, Persian, and Hebrew, as well as mathematics and astronomy.

10. Obit. S. Beever (HM 62832), HL.

11. Susanna Beever, *A Pocket Plea for Ragged and Industrial Schools: or, A Word for the Outcasts* (Edinburgh: Johnstone & Hunter, 1852); and *"Foodless, Friendless, in Our Streets"; Being a Letter about Ragged Schools addressed to Boys and Girls* (Edinburgh: Johnstone & Hunter, 1853). The journal was called *the Strines Journal*, and was described as "'a monthly magazine of literature, science, and art'" in John Heywood, ed., *Papers of the Manchester Literary Club*, vol. 22 (Manchester: Deansgate and Ridgefield, 1896), 410.

12. S. Beever, *King Lear; or, the Undutiful Children. A Tale in Twelve Chapters* (London: Bull, Simmons & Co., 1870). S. Beever, *A Book of References to Remarkable Passages in Shakespeare* (London: Bull, Simmons & Co., 1870). Beever's obituary also mentions that she published "a little book on the Bible in Shakspere [sic] . . . long ago" in Obit. S. Beever (HM 62832), HL. For "ill health":

obit. S. Beever (HM 62832); cf. S. Beever to William and Rosa Tuckwell, January 8, 1890 (HM 62893–62907) where she mentions she has suffered "cruelty" in her life.

13. Less than a thousand copies, see Clive Wilmer's introduction in John Ruskin, *Unto This Last and Other Writings*, ed. Clive Wilmer (New York: Penguin, 1985) 29; see also n. 37 below on why the book sold better later on. Rose La Touche: Hilton, *John Ruskin*, 487. Storm Cloud and illness: Hilton, *John Ruskin*, 491–92, and *Works*, 27.132ff.

14. *Hortus Inclusus* was subtitled *Messages from the Wood to the Garden*.

15. For greenhouses (hothouses): John Illingworth, "Ruskin and Gardening," *Garden History*, vol. 22, no. 2 (1994), 219–20; and *Works*, 28.181–84.

16. Illingworth, "Ruskin and Gardening," 222–23. William Robinson, *The Wild Garden* (London: John Murray, 1870).

17. On the suitability of various plants, we have relied on Richard Mabey, *Flora Britannica* (London: Chatto & Windus, 1997), and Clive Stace, *New Flora of the British Isles* (Cambridge, UK: Cambridge University Press, 2010). We also received helpful comments from Sally Beamish, Brantwood's estate manager, but all errors are our own.

18. For "*Gentiana verna*," see S. Beever to W. Tuckwell, October 25, 1889 (HM 62857–62892), HL; "*Potentilla fruticosa*" (HM 62915), HL. *Potentilla fruticosa* "is a relic of the last phases of the Ice Age" and 12,000 years ago grew in "glacial meltwaters" in England. Now it is found in just two sites, Upper Teesdale and the Lake District (see Mabey, *Flora Britannica*, 186). Stace, *New Flora*, 549, lists *Gentiana verna* as "Native" to the British Isles and "extremely local" in N. England and W. Ireland; as of September 13, 2013, http://www.cumbria-wildlife.org.uk/plants.html lists it as frequent on the "Cumbrian side of the Pennines."

19. S. Beever to W. Tuckwell (HM 62857–62892), HL: "rock roses" and "*lithospermums*," October 25, 1889; "rhododendrons," June 30, 1890; "Travellers Joy," October 24, 1890; "geraniums," February 21, 1891; "saxifrage," April 16, 1891; "sweet brier," October 23, 1891.

20. See S. Beever to W. Tuckwell (HM 62857–62892), HL: "*Schizostylus*" and "*Senecio pulcher*," October 25, 1889.

21. The friend was Dr. Guthrie. S. Beever to W. Tuckwell, n.d., ca. 1892 (HM 62857–62892),HL.

22. Poetical, obit. S. Beever (HM 62832), HL; cf. *Works*, 37.80. W. Tuckwell to unknown, n.d. (HM 62833), HL: intimate; Rob Roy (Beever's allusion was probably impish; Tuckwell says that she told Ruskin "she had always thought of him as Andrew [Fairservice] had described his master"; cf. *Works*, 34.295–97, where Ruskin gives this description in *Fiction, Fair and Foul*, along with analysis of Fairservice's complex, even deceitful character). Helen Gill Viljoen, *The Brantwood Diary of John Ruskin* (New Haven, CT: Yale University Press, 1971): all her life, 385; strange, 392.

23. Painting: W. G. Collingwood, "Miss Susan Beever at the Thwaite, 1892," Ruskin Museum, Coniston. See also "Ruskin Exhibition," *Daily News* (London), August 16, 1900, where only Beever's portrait is mentioned specifically. Owl of the Thwaite: S. Beever to W. Tuckwell, 29 May 1889 (HM62857–62892), HL, and also S. Beever to A. Fleming, where she signs her letters "Owletta" (HM 62834–62837), HL. Ghastliness: *Works*, 37.154, and Hayman, "John Ruskin's *Hortus Inclusus*," 368. Viljoen, *Brantwood Diary*: sparrows, 401; gooseberries, 384. "True old school": Ruskin to S. Beever, 1879 (JR289), HL.

24. Viljoen, *Brantwood Diary*: old-fashioned, officious, 399; wages, 402. Ruskin, *Hortus Inclusus*: like Thoreau,148; presumptuous, 156. Germination, *Works*, 37.202. Frightened to plant it, Ruskin to S. Beever, ca. 1883 (JR 395), HL. *Works*: specimen, 25.546; Latin, 37.201–2. Invaluable, S. Beever to W. Tuckwell, March 14, 1892 (HM 62857–62892), HL. Nasty gloxinias, *Works*, 37.369.

25. H. D. Rawnsley, *Ruskin and the English Lakes* (Glasgow: James MacLehose and Sons, 1902): lover, 35; impressed, 177. Childish feeling: *Works*, 5.369.

26. Jenifer Lloyd, "Raising Lilies: Ruskin and Women," *Journal of British Studies*, vol. 34, no. 3 (July 1995): guild women, useful members, 342; St. Ursula, 343.

27. S. Beever to W. Tuckwell (HM 62857–62892), HL: bantam cock (n.d.); weed with foot, grouse and marrows, April 16, 1891; peas, apples, pears, damsons, elderberries, melons, August 21, 1891; cabbages, April 21, 1890. *Works*: rosemary and lavender, 37.398; cranberries, 37.280; oranges, 37.289; brown bread, 37.290, 37.87, 37.363; asparagus, 37.363. Oranges did not grow well (presumably even in a greenhouse): Viljoen, *Brantwood Diary*, 389. For melons started indoors, S. Beever to W. Tuckwell, n.d., ca. spring 1891 (HM 62857–62892), HL.

28. J. Beever, *Practical Fly-Fishing*: fish pond and press, xvii–xviii; working men, 1–2. Thefts from gardens, Viljoen, *Brantwood Diary*, 398–99 (see p. 564 on the "Cousin Mary" who tells Beever about these thefts; it is not clear where the cousin lived exactly). Overworked man, S. Beever to W. Tuckwell, March 26, 1891 (HM 62857–62892), HL. Obit. S. Beever (HM 62832), HL.

29. Carving, J. Beever, *Practical Fly-Fishing*, xviii. Beever supported the Langdale Linen Industry, obit. S. Beever, HM 62832; also Sara Haslam, *John Ruskin and the Lakeland Arts Revival, 1880–1920* (Cardiff: Merton, 2004), 30. Beever learned to spin, C. H. Beever to S. Beever, November 6, 1884 (L65), RL; also Ruskin to S. Beever ca. 1885 (JR 489), HL. Inkstand, S. Beever to Ruskin, February 22, 1878 (L85), RL. Greatest in the least, *Works*, 37.290, italics ours; cf. "You . . . are so purely and brightly happy in all natural and simple things," 37.97. Feathers, *Works*, e.g., 37.74, 164, 180, 186, 220, etc.

30. For "cream cheese," see S. Beever to W. Tuckwell, n.d., ca. summer 1889 (HM 62857–62892), HL. Storyteller: J. Beever, *Practical Fly-Fishing*, xviii–xvix; Hayman, "John Ruskin's *Hortus Inclusus*," 368; see also S. Beever to W. and R. Tuckwell, January 27, 1891 (HM 62893–62907), HL (here she describes the cat Othello's adventure and a little book she thinks it could make). Dr. John Brown on Beever's letters, *Works*, 37.88.

31. Much sorrow, Viljoen, *Brantwood Diary*, 396. Idea for *Frondes Agrestes*, Hayman, "John Ruskin's *Hortus Inclusus*," 368. Pearls of wisdom, Christina Rieger, "'Sweet Order and Arrangement': Victorian Women Edit John Ruskin," *Journal of Victorian Culture*, vol. 6, no. 2, 231–49 (231–32, 238).

32. *Works*: form, cut out preaching, didactic, value, 37.108; wholly, in sympathy, 37.117; *Word Painting*, 37.136. Agrestes, Viljoen, *Brantwood Diary*, 382; but cf. Hayman, "John Ruskin's *Hortus Inclusus*," 369, where Hayman says Ruskin translated *Frondes Agrestes*, "The wood sheds for thee its wild leaves." Sugary stuff, Hayman, "John Ruskin's *Hortus Inclusus*," 369. John Ruskin and S. Beever, eds., *Frondes Agrestes: Readings in Modern Painters* (Sunnyside: George Allen 1875): turning out right, 149; useful, vi. Not meant for girls, *Works*, 37.338–39.

33. Ruskin and S. Beever, eds., *Frondes Agrestes*: Caligula like, 4–5 (cf. *Works*, 4.61–62); rest, 159 (cf. *Works*, 4.114); physical exertion, 143 (cf. *Works* 7.428–29); evil nature, 140 (cf. *Works*, 7.426, where it says "nation" instead of "nature"); art of joy, 140–41, italics ours (cf. *Works*, 7.427).

34. Rieger, "'Sweet Order and Arrangement,'" 245. See Louisa Tuthill, ed., *The True and the Beautiful: Nature, Art, Morals, and Religion* (New York: John Wiley & Sons, 1890); Rose Porter, ed., *Nature Studies: Selections from the Writings of John Ruskin* (Boston: Dana Estes & Co., 1900).

35. Absolute submission: Ruskin, *Frondes Agrestes*, vi. 1,000 copies, Wilmer, *Unto This Last*, 29. V.A.: I am grateful to the anonymous reviewer for *the Eighth Lamp* for pointing out that the publisher's management of the first edition may have affected sales negatively, and that in 1877 *UTL* was reissued and began to sell very well under George Allen. See Paul Dawson, *George Allen of Sunnyside; To accompany an exhibition celebrating the centenary year of George Allen, 1832–1907*, with contributions by Stephen Wildman (Lancaster: Ruskin Library, 2007), 16. Further, Ruskin had in mind that *Frondes* would be a "*cheap*" book (*Works*, 37.117), which would have helped sales too. That said, Beever's selections may have gently prepared some readers for a fuller argument regarding sufficiency prior to the reissue of *Unto This Last*. 34,000 copies, Hayman, "John Ruskin's *Hortus Inclusus*," 369. Ð70 a year, Collingwood to "Lammie," July 26, 1893, AH.

36. Hyères, S. Beever to W. Tuckwell, February 21, 1891 (HM 62857–62892), HL. Elsewhere she says her "table is full of flowers some from Hyères," though it is unclear whether they were ordered or received as gifts: n.d., ca. 1892 (HM 62857–62892), HL. Beever also mentions that she

expects to receive rose trees from the bishop of Brechin, who "gets his roses from Lyons." They are mentioned directly after the bishop's thank-you letter for some "seed of Shirley Poppies" she sent him. The rose trees are thus a symbol of friendly exchange. See S. Beever to R. Tuckwell, October 27, 1890 (HM 62838–62856), HL. American garden, *Preston Guardian*, October 7, 1848. Note that she almost sent to London for port once, but Ruskin gave it to her instead: Ruskin, *Works*, 37.285. Did not travel, Ruskin, *Hortus Inclusus*, v (but note that Beever's obituary mentions that her family traveled to Scotland at times, S. Beever, obit, HL).

37. Against animal abuse: obit. S. Beever (HM 62832), HL. Devilish, S. Beever to A. Fleming, n.d., ca. 1887 (JR 555–753), HL. Butter, John Ruskin, *Hortus Inclusus*, 158. No one in Brantwood, Viljoen, *Brantwood Diary*, 402. Cometh not, S. Beever to W. and R. Tuckwell, March 6, 1891 (HM 62897), HL. Painfully nervous, S. Beever to A. Fleming, September 10, 1887 (JR 555–753), HL. Avoiding, Viljoen, *Brantwood Diary*, 402.

38. *Works*: too modest, 37.630; centre of order, 18.137; public work, 18.136. Koven, "How the Victorians Read *Sesame and Lilies*": unstable ideology, 167; Octavia Hill, 179–81. Also on women's "wider public duties," see Lloyd, "Raising Lilies: Ruskin and Women," 340, 345. Effie Gray, Hilton, *John Ruskin*, 197.

39. Leonore Davidoff and Catherine Hall, *Family Fortunes: Men and Women of the English Middle Class 1780–1850* (New York: Routledge, 2002), 429–36. S. Beever, *A Pocket Plea for Ragged and Industrial Schools*, 11; also see n. 13 above.

40. 1882 edition, aged friend, see Ruskin, D. Nord, ed., *Sesame and Lilies*, 23. "Queen Susan," *Works*, 37.411. Inhuman labor, Koven, "How the Victorians Read *Sesame and Lilies*," 173. Coal shaft, *Works*, 18.133–34.

41. *All* done for us, other creatures, Ruskin, *Frondes Agrestes*, 36. Plague-wind, Works, 27.132–33; also Hilton, *John Ruskin*, 492. Change of climate, *Works*, 37.101. "London-best" fog, Viljoen, *Brantwood Diary*, 382. Brown sponges, *Works*, 34.38. Viljoen, *Brantwood Diary*: deposit from Manchester, 201; dimming sun, 146..

42. Mother earth, Ruskin, *Hortus Inclusus*, 153. Ashamed, Ruskin to S. Beever, ca. May 1875 (JR 1–526), HL, This July, Ruskin to S. Beever, July, ca. 1878–80 (JR 239), HL.

43. Gloom, S. Beever to Ruskin, ca.1875, in Viljoen, *Brantwood Diary*, 402. Heaped ashes, Ruskin to S. Beever, 1878–82 (JR 278), HL. S. Beever to A. Fleming (JR 555–753), HL: tempestuous, n.d., ca. 1887; shrewdly, "Thursday," n.d., ca. 1887; Brantwood weather, n.d., ca. 1887.

44. Animals, obit. S. Beever (HM 62832), HL. For "vivisection" and "animal welfare," see Hilton, *John Ruskin*, 792, 794. S. Beever to A. Fleming (JR 555–753), HL: cruel, May 19, 1887 (that it was Ruskin who gave away the pony is deduced from many letters before and after this particular one); greatest pleasures, September 26, 1887.

45. *Hobby Horse*, 1,700 copies, Hayman, "John Ruskin's *Hortus Inclusus*," 374–75 (1,700 combines the large- and small-edition sales). City on a hill, *Works*, 37.327.

46. Brechin, S. Beever to R. Tuckwell, October 27, 1890 (HM 62838–62856), HL, and S. Beever to W. and R. Tuckwell, January 30, 1891 (HM 62896), HL. American ladies, S. Beever to W. Tuckwell, n.d., ca. 1889 (HM 62857–62892), HL; "rosebud": S. Beever to A. Fleming, October 11, 1887 (JR 555–753), HL. Cardinal Manning, see S. Beever to A. Fleming, October 5, 1887 (JR 555–753), HL. William Tuckwell, *Tongues in Trees and Sermons in Stones* (Orpington: George Allen, 1891), 113.

47. Ibid., 112.

48. S. Beever to A. Fleming (HM 62834–62837), HL: "season," n.d.; "starved," n.d. Wallace, March 14, 1890 (HM 62857–62892), HL. She also mentions *The [Voice] of a Naturalist*, July 13, 1890 (HM 62857–62892), HL, and *The Naturalist at Home*, S. Beever to W. and R. Tuckwell, January 8, 1890 (HM 62893–62907), HL. However, she might have meant Biot Edmondston and Jessie Margaret Edmondston, *The Home of a Naturalist* (London: James Nisbet & Co. 1889). Thick and heavy, S. Beever to W. Tuckwell, n.d. (HM 62857–62892), HL. Lapland, S. Beever to W. Tuckwell, October 23, 1891 (HM 62857–62892), HL. William Tuckwell, *Tongues in Trees*: rare plants, 112; paradise, 113.

49. Sessions, *Literary Celebrities*, 187–93. *The Garden: An Illustrated Weekly Journal of Gardening in All Its Branches*, October 1, 1921, vol. 85, 495. This journal was founded by William Robinson. Prayed, neglected, Bruce Hanson, *Brantwood: John Ruskin's Home 1872–1900*. Brantwood Trust, Cumbria [pamphlet, n.d.], 28. David Ingram, *The Gardens at Brantwood: Evolution of John Ruskin's Lakeland Paradise* (Pallas Athene & the Ruskin Foundation, 2014), 48, 74.

50. We are grateful to the former owners of Thwaite Cottage, Marguerite and Graham Aldridge, for sharing with us the last known map of Beever's garden and indicating how they think things may have been arranged. Thrones, *Works*, 37.494.

51. Woman's voice, S. Beever to W. Tuckwell, n.d. (HM 62857–62892), HL.

52. Elizabeth Harcourt Rolls Mitchell, *First Fruits* (London: Hurst and Blackett, 1857). The poem is called "Watchman! What of the Night?"

CHAPTER FOUR

1. L. T. C. Rolt, *Red for Danger: A History of Railway Accidents and Railway Safety Precautions* (London: Bodley Head, 1955), 63–64.

2. *Works*: unseemly rags, 28.298, 300; gigantic, 28.147; cp. 27.664–66; luggage trains (August 1879), 34.37.

3. Robert Somervell, *A Protest against the Extension of Railways of the Lake District* (London, 1876), 1, 7–8; cf. James M. Garrett, *Wordsworth and the Writing of the Nation* (Farnham, UK: Ashgate, 2013), 149, 182.

4. *Oxford Dictionary of National Biography*, "Rawnsley, Hardwicke Drummond (1851–1920)"; Wheeler, *Ruskin and the Environment*, 157.

5. For the "steam dragon," see H. D. Rawnsley, *Ruskin and the English Lakes* (Glasgow: James MacLehose, 1902), 41; idem, "The Proposed Permanent Lake District Defence Society," in *Transactions of the Cumberland Association for the Advancement of Literature and Science: Part VIII, 1882–83*, ed. J. G. Goodchild (G. Carlisle & T. Coward, 1883), 69.

6. Richard Bradley and Mark Edmonds, *Interpreting the Axe Trade: Production and Exchange in Neolithic Britain* (Cambridge, UK: Cambridge University Press, 1993); M. L. Ryder, *Sheep and Man* (London: Duckworth, 1983), 192; G. P. Jones, *The Lake Counties, 1500–1830: A Social and Economic History* (Manchester, UK: Manchester University Press, 1961); I. G. Simmons, *An Environmental History of Great Britain* (Edinburgh: Edinburgh University Press, 2006[2001]), 105–7; J. E. Williams, "Whitehaven in the Eighteenth Century," *Economic History Review*, New Series, vol. 8, no. 3 (1956), 398.

7. T. C. Smout, *Nature Contested: Environmental History in Scotland and Northern England Since 1600* (Edinburgh: Edinburgh University Press, 2000), 48.

8. "Michael," William Wordsworth, *Lyrical Ballads, 1798 and 1802*, ed. Fiona Stafford (Oxford: Oxford University Press, 2013), 285–97; Annabel Patterson, *Pastoral and Ideology: Virgil to Valéry* (Berkeley: University of California Press, 1988), 274, 279–80.

9. Somervell, *Protest*, 7, 27–28. *Works*: border peasantry, 34.141; esplanade, 34.569. Jonathan Bate, *Romantic Ecology: Wordsworth and the Environmental Tradition* (London: Routledge, 1991), 47–58; Scott Hess, *William Wordsworth and the Ecology of Authorship* (Charlottesville: University of Virginia Press, 2012), chapter 4, passim.

10. Ritvo, *The Dawn of Green*, 23, 81; women's voices, Ruskin to Somervell, Brantwood 23rd n.m. 1875, RL, L114 (5).

11. Ritvo, *The Dawn of Green*, 45, 62, 68–70, 81.

12. *Works*, 29.224; J. D. Marshall, *The Lake Counties*, 209; Ritvo, *The Dawn of Green*, 16–18, 82; "Are We to Preserve Our English Lakes?" *Daily News* (London), June 12, 1877; Ian Thompson, *The English Lakes: A History* (London: Bloomsbury, 2010), 233; Charles-François Mathis, *In Nature We Trust: Les paysages anglais à l'ère industrielle* (Paris: Librarie pups, 2010), 291–96; "The Proposed Improvement of Thirlmere," *Pall Mall Gazette*, London, November 8, 1877; "The Thirlmere Water

Works," *Standard*, London, November 2, 1877, p. 2; John Hammond Fowler, "Correspondence: Thirlmere and Manchester," *York Herald*, York, November 3, 1877, p. 6.

13. Ritvo, *The Dawn of Green*, 118; Graham Murphy, *Founders of the National Trust* (London: National Trust, 2007); Ian Thompson, *The English Lakes*, 255.

14. *Bristol Mercury and Daily Post*, February 11, 1878.

15. Ibid.; Ian Thompson, *The English Lakes*, 246; clever hands, Rawnsley, *Ruskin and the English Lakes*, 117.

16. Rawnsley, *Ruskin and the English Lakes*, 116, 118-19; cf. Haslam, *John Ruskin and the Lakeland Arts Revival, 1880-1920* (Cardiff: Merton, 2004), 87-88.

17. Several printed versions of the KSIA's sale list are on file at CACC. No dates are given, however. The pamphlets are all titled "List of Articles made at the Keswick School of Industrial Arts."

18. Showroom photo, Haslam, *Lakeland Arts*, plate 8.

19. Advice, Haslam, *Lakeland Arts*, 92. Rawnsley, *Ruskin and the English Lakes*, 122-24, 127, 129.

20. Haslam, *Lakeland Arts*: Keswick school, 198; speeches, 199; 1891 exhibition, 200-201 (businesses could "submit locally made products"); hoveldom, 203-4; Carlisle, 206-7.

21. Michael Dowthwaite, "Defenders of Lakeland: The Lake District Defence Society in the Late-Nineteenth Century," in Oliver Westall, *Windermere in the Nineteenth Century* (Lancaster: Lancaster University Press, 1991), 50, 52, 55; Thompson, *The English Lakes*, 248.

22. For Rawnsley and telephone lines, see CACC DSO 24/4; see also Haslam, *Lakeland Arts*, 100; "St. George's knights," quoted in John K. Walton, "The National Trust: Preservation or Provision?" in Wheeler, *Ruskin and the Environment*, 156-57; Thompson, *The English Lakes*, 253-55.

23. "Dalesmen," "not safest guardians," H. D. Rawnsley, "The Proposed Permanent Lake District Defence Society," 1883, CACK WDX 422/2/3, p. 10; "severely simple life," H. D. Rawnsley, *Lake Country Sketches* (Glasgow: James MacLehose and Sons, 1903), 4; "tradesmen's word," *Works*, 34.142.

24. Rawnsley, *Lake Country Sketches*, 3-4.

25. Rawnsley, *Lake Country Sketches*: white, 9; not inventive, 5; unimaginative, 6; simple life, 8; cliverish, 24; Dorothy, 35; Westmoreland gentleman, 16-17; shy, 18-19; fwoaks, 38; nobbut a stean, 9-10; Wordsworth and wife, 45, 57; not for money, 58; needed help, 58.

26. Rawnsley, *Lake Country Sketches*: seclusion, blindness, forgot faults, 4-5; Hartley Coleridge favored, 23-24; public-houses, 13.

27. Rawnsley, *Ruskin and the English Lakes*: Wordsworth aloof, 169; Ruskin's approachableness, 169-70; lace, 135. "Ruskin Lace" was based on an intricate Greek lace design.

28. Generation to generation, H. D. Rawnsley, "Our Pulpit, The Temporal and the Eternal: A Rushbearing Sermon," August 14, 1908 (newspaper clipping), CACK, WDX 402/10. The file also contains many other Rushbearing sermons given between 1908 and 1918. See also H. D. Rawnsley, *Life and Nature at the English Lakes* (Glasgow: James MacLehose and Sons, 1902), 1-16.

29. H. D. Rawnsley, "A Rushbearing Sermon, St. Theobald's Church," July 6, 1911, CACK, WDX 402/10, pp. 3, 5. Simple labor, farmer, H. D. Rawnsley, "A Harvest Sermon: The Dignity of Manual Labour," CACK, WDX 402/3/8, pp. 17-19.

30. Poet, Goths, "Thirlmere Today," *Manchester Times*, October 19, 1894; Robert Somervell, *Chapters of Autobiography* (London: Faber & Faber, 1935), 50-56; party of innkeepers, Ritvo, *The Dawn of Green*, 133.

31. Uncanny, bleak, nobler, "Thirlmere Today," *Manchester Times*, October 19, 1894. "Too deep," quoted in Ritvo, *The Dawn of Green*, 141. Cairn, H. D. Rawnsley, *Literary Associations of the English Lakes*, vol. 2 (Glasgow: James MacLehose & Sons, 1906, 3rd ed.), 225, 227-28. Thirst and strife, "To the Workers," happy sonnets, "Thirlmere Today," *Manchester Times*, October 19, 1894.

32. H. D. Rawnsley, "Sunlight or Smoke?" *Contemporary Review*, vol. 57 (1890: January/June), 512-13.

33. Rawnsley, "Sunlight or Smoke?" 514–16; Herbert Fletcher, "On 'Smoke Abatement,'" *Transactions of the Sanitary Institute of Great Britain*, vol. 9 (London 1888), 303–8; Thorsheim, *The Invention of Pollution*, 101.

34. Rawnsley, "Sunlight or Smoke?": factories visited, 513, 517, 518, 519. These also included Wardle and Brown's weaving mills as well as those of Messrs. Crosses and Winkworth, Atlas Mills of Messrs. Musgrave & Co., Ltd., and Messrs. Bayley & Sons, P. Crook & Co., Ltd., and Canon Brothers; public spirit, 519; quick vs. slow, 520; no public conscience, 521; Sunlight Soap, 524; their wealth, 524.

35. *Works*, 27.182–83, 28.129–30.

36. H. D. Rawnsley, *Ruskin and the English Lakes*, 79–83. Wolfgang Schivelbusch suggests a contrast between the preindustrial "intensive" gaze and the panoramic perception of modern train travelers. *The Railway Journey: The Industrialization in Time and Space in the 19th Century*, 3rd ed. (Berkeley: University of California Press, 1986), 57, 189.

CHAPTER FIVE

1. W. G. Collingwood, *Thorstein of the Mere* (Kendal: T. Wilson, 1895), 262.

2. WGC to EMDC, AH: Liverpool illness, August 2, 1883; headache, December 16, 1893; seaside resorts (New Brighton), March 4, 1883. Cf. Peter Thorsheim, *The Invention of Pollution: Coal, Smoke, and Culture in Britain since 1800* (Athens: Ohio University Press, 2006).

3. Best and dearest, *Works*, 26.568. Physical description of WGC, Cumberland and Westmoreland Antiquarian and Archaeological Society obituary for WGC, WDWGC/3/9, p. 311, CACK. W. G. Collingwood, *The Philosophy of Ornament: Eight Lectures on the History of Decorative Art* (Orpington: George Allen, 1883), cf. 209–10.

4. Matthew Townend, *The Vikings and Victorian Lakeland: The Norse Medievalism of W. G. Collingwood and His Contemporaries* (Kendal: Titus Wilson & Son, 2009, for Cumberland and Westmoreland Antiquarian and Archeological Society), 90. Loom, April 11 and 18, 1883, WGC to EMDC (WGC/386/15, 17), AH.

5. This interpretation comes quite close to the argument made by Karen Welberry about Collingwood's "use of Ruskin in a domestic setting"; see "H. D. Rawnsley, W. G. Collingwood, and the German Miners of Keswick 1565–1645: An Early Quarrel over Conservationist Paradigms in Lakeland," in *Collingwood and British Idealism Studies*, eds. D. Boucher, B. Haddock, A. Vincent (Cardiff: R. G. Collingwood Society, 2004), vol. 1.10, 6.

6. WGC to EMDC (WGC/386/27, 29—transcriptions), AH: no good, August 2, 1883; living at all! August 8, 1883; Beever's gift, August 10 and 23, 1883, and cf. October 25, 1883.

7. See Robert and Edward Skidelsky, *How Much Is Enough: Money and the Good Life* (New York: Other Press, 2012), 3.

8. Detailed letter, WGC to EMDC, AH, October 19, 1883. Childhood nurse, Townend, *The Vikings and Victorian Lakeland*, 29.

9. WGC to EMDC, AH: solemnity, October 19, 1883; childishness, October 25, 1883.

10. WGC to EMDC, AH: rotten house, dialect, loneliness, October 28, 1883; safer man, October 25, 1883.

11. WGC to EMDC, AH: gloriousness, reach hills, October 28, 1883; independent, October 25, 1883.

12. Quoted in Townend, *The Vikings and Victorian Lakeland*, 30.

13. Purgatory, Townend, *The Vikings and Victorian Lakeland*, 32. Can't paint, WGC to EMDC, no date [1889], AH (as cited by Townend, 36–37).

14. John Ruskin (ed.), Alexander Wedderburn and W. G. Collingwood (trans.), *The Economist of Xenophon* (London: Ellis and White, 1876); W. G. Collingwood, *Ruskin Relics* (New York: T. Y. Crowell & Co., 1904), 9.

15. Collingwood, *The Philosophy of Ornament (1883)*: civilization hinders, 4–5; by hand, 208; industrialization, 209–10.

16. Pleasure of it, Sara Haslam, *John Ruskin and the Lakeland Arts Revival, 1880–1920* (Cardiff: Merton, 2004), 137. Much wants more, WGC to Arthur W. Simpson, June 22, 1886, CACK.

17. Beever's additional gift, WGC to his father, August 2, 1887, AH; Teresa Smith emphasizes the difficult circumstances too in "R. G. Collingwood's Childhood: Habits of Thought," in R. G. Collingwood, *R. G. Collingwood: An Autobiography and Other Writings with Essays on Collingwood's Life and Work* (Oxford: Oxford University Press, 2013), 194.

18. The quarterly rent and WGC's quarterly salary were £25. Townend, *The Vikings and Victorian Lakeland*, 43–44, esp. 44, n. 123.

19. CACK: amateurish, WGC to AWS, Wed. 7, n.y. (WDX515/2/1/2/57); sofa, WGC to AWS, May 22, 1891 (WDX515/2/1/2/55); floor, WGC to AWS, July 24, 1891 (WDX515/2/1/2/62).

20. CACK: settle, etc., WGC to AWS, October 27, 1891 (WDX515/2/1/2/67); see also July 24, 1891 (WDX515/2/1/2/62).

21. *Works*: Ruskin's furniture, 4.7–8, n. 2; nothing allowed to remain, 1.14–15. On the rise of the antiques trade, see Deborah Cohen, *Household Gods* (New Haven, CT: Yale University Press, 2006), chapter 6, "Designs on the Past: Antiques as Faith."

22. Lawn, WGC to EMDC, July 5, 1894, AH.

23. CACK: art market, WGC to "Sir" (perhaps Howard Day), April 20, 1895 (WDX/653/3); Simpson's prospectus, WGC to AWS Monday n.d. [but in a different hand is added: 17.IV.93] (WDX515/2/1/2/72).

24. Spoony, WGC to EMDC, August 20, 1893, AH. Private means, Townend, *The Vikings and Victorian Lakeland*, 228.

25. Best Home Made, Ursula to WGC and EMDC, March 22, 1903 (WDWGC/1/6), CACK.

26. Speculation, WGC to EMDC, November 1[7], 1894, AH. Dora's expenses, n.d., WDWGC/1/5, CACK (the record is kept on scratch paper, on the back of which is printed University College, Reading, Dept. of Fine Arts, 1899). Sold picture, Ursula to EMDC, October 3, 1907 (WDWGC/1/6), CACK.

27. WGC to EMDC, AH: sick children, August 14 and 15, 1893; visitors, Ruthie, August 8, 1893; donkey, July 20, 1893, and August 4, 1893. They may have had other servants as well.

28. Townend, *The Vikings and Victorian Lakeland*: Emma Holt, 231; Reading, 231; job paid for education, 250; salary, 231; CWAAS, 228; trifling return, 252; subscriptions, 211.

29. A 1928 letter from Ewart James to Rawnsley's wife, Edith, reveals that Collingwood's more subtle efforts did not go unnoticed. James wrote about the need for a consolidated preservationist association in the Lake District but noted that several individual groups had "done" and were "doing much," including "The C&W Antiquarian Society (under Mr. Collingwood)": Ewart James to Mrs. Rawnsley, November 19, 1928, p. 10 (WDX/422/2/3), CACK.

30. Townend, *The Vikings and Victorian Lakeland*, 147; Marwick comment, NLS 2984, 81–82.

31. Townend, *The Vikings and Victorian Lakeland*: portraits, amiable, 35; potboilers, 251. Flourishing, WGC to his father, June 13, 1893, AH.

32. Ritvo, *The Dawn of Green*, 127–28; W. G. Collingwood, *Thorstein of the Mere: A Saga of the Northmen in Lakeland* (Kendal: T. Wilson, 1895), 78; Townend, *The Vikings and Victorian Lakeland*, 66.

33. For the failed battle for the preservation of Thirlmere, see chapter 4, as well as Ritvo, *The Dawn of Green*. W. G. Collingwood, *The Lake Counties* (London: J. M. Dent and Sons Ltd., 1949, based on rev. ed., 1932), 153.

34. W. G. Collingwood, *Thorstein of the Mere: A Saga of the Northmen in Lakeland*, unbuilt and uncleared, 83. Townend, *The Vikings and Victorian Lakeland*, on composition of *Thorstein*, 55–63, multilingual, 66.

35. Collingwood, *Thorstein of the Mere*: not idle, 65; "daintiness," 66; York, 245–50.

36. Collingwood, *Thorstein of the Mere*: Asdis idle, 225; Raineach weeping, 98; Raineach courageous, 140, 278; snug home, 265; holiday, 266.

37. Independence, Collingwood, *The Lake Counties* (London: J. M. Dent & Co., 1902), 158–59. Birthright, Collingwood, *Thorstein of the Mere*, 303–4.

38. For an in-depth account of Collingwood's journey to Iceland and more on the idea that he hoped to find "that Iceland and Lakeland could be regarded as parallel colonies, settled in the same period, as part of the same historical process, by migrants from mainland Scandinavia," see Townend, *The Vikings and Victorian Lakeland*, 87ff. On the business scheme, see ibid., 106.

39. WGC to daughter Barbara, June 8, 1897, in W. G. Collingwood, *Letters from Iceland* (Swansea: R. G. Collingwood Society, 1996), 12–13.

40. Collingwood, *Letters from Iceland*: poor and mean, 15; masons, no costume, neglect, 16; penny paints, 19; deadly ugly, 21.

41. Collingwood, *Letters from Iceland*: true stories, 44; drew it as it was, 45. Townend, *The Vikings and Victorian Lakeland*, 111. Townend notes that Collingwood was not the first to paint these scenes, but was likely unaware of Sabine Baring-Gould's *Iceland: Its Scenes and Sagas* (1863). Union of story and scenery, Collingwood, *The Lake Counties* (London: J. M. Dent & Co., 1949), 4. Collingwood also repeats here Ruskin's story about a horrifying revelation he had in the Alps when he attempted to imagine the land as if it had never been inhabited, never once, by any human being: "'a sudden blankness and chill were cast upon [the land]; the flowers in an instant lost their light, the river its music; the hills became oppressively desolate.'"

42. Collingwood, *Letters from Iceland*: prejudiced, 36; "Saurar—Sawrey," 39.

43. Spoil-heaps, WGC to Gordon Wordsworth, October 3, 1924 (WDX 422/2/3), CACK (NB: Collingwood speaks here of heaps on Spionkopf, his private name for the Old Man of Coniston).

44. Collingwood, *Ruskin Relics* (1904), 56.

45. Collingwood, *Life and Works of John Ruskin* (London: Methuen & Company, 1893, 2 vols.), vol. 2, 237–38. W. G. Collingwood, *The Lake Counties* (London: J. M. Dent & Co., 1902): "grass," 66; other references to smoke and pollution, 74, 78, 85, 116, 133; *The Lake Counties* (1949), 95. *Ruskin Relics* (1904), utterances, 55, coronation night, 56, perfectly accurate, 59. In the 1911 edition of the Ruskin biography, Collingwood inserted a new line in the chapter on the Storm Cloud that defined it as the product of "the Smoke Nuisance." *Life of John Ruskin* (London: Methuen, 1911, 7th ed.), 290.

46. Peter Thorsheim, *Inventing Pollution*: smoke-abatement initiatives and education, 93–94; exhibition, 95–98; quantify pollution, concentration peaked (Peter Brimblecombe), 128. Hardwicke Rawnsley, "Sunlight or Smoke?" *Contemporary Review*, vol. 57 (1890, January/June), 512–24. John Graham, *The Destruction of Daylight* (London: George Allen, 1907): Rollo Russell, Lake District, foundries, 13–14; Christmas fog, 25; £1 per head, 27; Storm Cloud, 31–33. Collingwood, *The Lake Counties* (1902), 66. Something of this technical optimism may be found in Ruskin too; see the passage on our power to vitiate or *purify* air, water, and earth in *Fors Clavigera*, letter 5, *Works*, 27.91.

47. Collingwood, *Letters from Iceland*, 46.

48. Collingwood, *The Lake Counties* (1949): iron-smelting, 46; *vis medicatrix*, 61. W. G. Collingwood, *The Book of Coniston* (Kendal: T. Wilson, 1897): sledge road, riving of blocks, 7; black country, 21.

49. H. D. Rawnsley, "Sunlight or Smoke?" *Contemporary Review*, vol. 57 (1890: January/June). Karen Welberry, "H. D. Rawnsley, W. G. Collingwood, and the German Miners of Keswick 1565–1645: An Early Quarrel over Conservationist Paradigms in Lakeland," *Collingwood and British Idealism Studies*, vol. 10, 1–22.

50. Collingwood, *The Lake Counties* (1949): Romans, xvi; fox, 95.

CHAPTER SIX

1. Description of Lanehead's interior: Taqui Altounyan, *In Aleppo Once* (London: John Murray, 1969), 174 (items are recollected from visits in the 1920s and 1930s). Oil paint: Taqui Al-

tounyan, *Chimes from a Wooden Bell: A Hundred Years in the Life of a Euro-Armenian Family* (London: I. B. Tauris & Co., 1990), 57. Furniture: WGC to Arthur Simpson (AWS), July 24, 1891 (WDX515/2/1/2/62) and October 27, 1891 (WDX515/2/1/2/67), CACK. Daily routine: R. G. Collingwood, *R. G. Collingwood: An Autobiography and Other Writings with Essays on Collingwood's Life and Work* (Oxford: Oxford University Press, 2013), 3; Fred Inglis, *History Man: The Life of R. G. Collingwood* (Princeton, NJ: Princeton University Press, 2009), 4–5. See also Arthur Ransome, ed. Rupert Hart-Davis, *The Autobiography of Arthur Ransome* (London: Jonathan Cape, 1976), 94–95. There is a faded photograph of Lanehead's "Morning Room" included in *What Ho!* August 1905, AH.

2. Cry, WGC to EMDC, Monday, n.d., ca. 1883, from Haslemere, AH. The home was Gillhead on Lake Windermere, not (yet) Lanehead, but the important point was the landscape. Superancestor, Altounyan, *Aleppo*, 10.

3. See for instance Teresa Michals, "Experiments Before Breakfast: Toys, Education, and Middle-Class Childhood," in *The Nineteenth-Century Child and Consumer Culture*, ed. Dennis Denisoff (Hampshire: Ashgate, 2008): toys, commercialism, and middle-class anxiety, 29; children as consumers, 31; doting parents, 36. Ads targeting children: Lori Anne Loeb, *Consuming Angels: Advertising and Victorian Women* (Oxford: Oxford University Press, 1994), 30.

4. Talking posh, Inglis, *History Man*, 6. We are unsure of Inglis's source for this description, but suspect it was one of the Collingwood descendants.

5. Papier-mâché, sagas, pumps: R. G. Collingwood, *R. G. Collingwood: An Autobiography and Other Writings with Essays on Collingwood's Life and Work* (Oxford: Oxford University Press, 2013), 1; also Inglis, *History Man*, 5. Egalitarianism: Admittedly, wealthy families sometimes educated their daughters merely because they were educating a son at home anyway (Carol Dyhouse, *Girls Growing Up in Late Victorian and Edwardian England* [London: Routledge, 1981], 40–41), but Collingwood arguably promoted a more than typically independent-minded, broad, and rigorous curriculum for his daughters. Taqui Altounyan presents contradictory statements, saying that for WGC the girls' external education "did not matter," but also noting Dora's appreciation for the excellent education she received at home: *Chimes*, 43, 41. Though all the children were educated at home, Robin later attended Rugby, then Oxford. Ursula attended a boarding school as well. Dora and Barbara both attended art school in London and then University College, Reading (Matthew Townend, *The Vikings and Victorian Lakeland*, CWAAS, 2009, 231).

6. Daughters read *The Ethics of the Dust*, Dora to "Molly" (EMDC), ca. 1899 (WDWGC/1/5), CACK. Works: fight furiously, 18.278; work and play, nobler life, activity of hope, 18.360; enact structures, 18.235ff; mica, 18.253; diamonds, 18.217; happy, 18.296; careless production (see the "unhappy metal worker"), 18.244.

7. AH: monthly, *NM*, January 1897; subscribers, circulation, *What Ho!* August 1905 (we assume the subscribers to *Nothing Much* were similar based on editorial comments and the variety of guest contributors in *NM*); future days, *NM*, New Year 1903; cribs, *NM*, December 1901. H. H. Warner, ed., *Songs of the Spindle & Legends of the Loom* (London: N. J. Powell & Co., 1889). On other adolescent newspapers and magazines, see Jessica Isaac, "Youthful Enterprises: Amateur Newspapers and the Pre-History of Adolescence, 1867–1883," *American Periodicals: A Journal of History, Criticism, and Bibliography*, vol. 22, no. 2 (2012), 158–77. The American youths described by Isaac printed multiple copies of their editions and included advertising from local businesses.

8. Sara Atwood, *Ruskin's Educational Ideals* (Surrey: Ashgate, 2011): active learning, 119; spin, cook, etc., 142; see also *Works*, 27.143. Sailing: Note that the Collingwoods descended from an impressive line of naval officers (Inglis, *History Man*, 4). Run and find out, *NM*, April 1902. That Ruskin insisted one should find out for oneself rather than be told: Atwood, *Ruskin's Educational Ideals*, 102. Monikers, *NM*, *passim*, AH.

9. AH: Barrow smoke, *NM*, March 1900; Coniston smoke, *NM*, September/October 1902; storm cloud photo, *What Ho!* August 1905.

10. AH: poem, WGC to "Budie" (Dorrie), October 16, 1887; chaffinch, *NM*, April 1898; siskin,

NM, July 1898; egg, *NM*, June 1902. Only the neighbor's moniker is given, "Professor Bombastes M. M. M. C."

11. AH: three planets, *NM*, December 1901; lair, *What Ho!* August 1905 (see also Ransome, *Autobiography*, 96). On "seeing" as a precursor to deeper moral and intellectual understanding, see Atwood, *Ruskin's Educational Ideals*, 54. See also R. G. Collingwood, *Essays in the Philosophy of Art*, ed. Alan Donagan (Bloomington: Indiana University Press, 1964), xii (where he describes Robin's [R. G. Collingwood's] adult estimation of this sense of "seeing"). *Works*: worthy of the soil, 20.37; pollute, 20.36; see also Atwood, *Ruskin's Educational Ideals*, 63.

12. Pitsteads, *NM*, December 1901, AH.

13. AH: Atkinson ground, water-course, wall, *NM*, August 1901; triangular holes, *NM*, April 1902.

14. AH: luckystones, bullets, *NM*, February 1901; dreadnoughts, *NM*, September 1898 . . . for other warfare technology, see for instance *NM*, May 1902, on cat warfare in "The Customs of Cat Land"; blast, *NM*, June 1902.

15. Ruskin on ads, *Works*, 27.49. WGC on ads: WGC to Dorrie, January 18, 1883, AH; also WGC to William Marwick, October 10, 1889, NLS 2984, f. 48.

16. AH: cocoa ads, *NM*, March 1900; Locklick's, *NM*, March 1897; Kirkbride's, *NM*, July 1897.

17. AH: Xmas ad, *NM*, October 1899; Queen's visit, *NM*, December 1899.

18. AH: presents, WGC to "Budie" (EMDC), February 21, 1894; big printing press, etc., *NM*, January 1898; doll store, *NM*, July 1897. CACK: playing shops, Dora to "Molly" (EMDC), January 21, 1898 (WDWGC/1/3); children gardening, Dora to WGC and EMDC, 1896, 1898 (WDWGC/1/5); basket for kittens, Dora to Robin, "Dear Demon," WDWGC/2/3; bookbinding, Ursula to WGC and EMDC, 1903 (WDWGC/1/6).

19. AH: monthlong stay, WGC to father, August 2, 1887; tea, *NM*, December 1900; oeuf, *NM*, February 1902. Mines, SB to Fleming, "Friday" n.d., ca. 1887 (JR 555–753), HL. Generations, Altounyan, *Aleppo*, 179.

20. Barbarians, *NM*, May 1899, AH.

21. Twentieth century, Mainz, *NM*, June 1900, AH.

22. "On the Nature of the Gothic," *Works*, 10.180ff. Critique of political economy, Ruskin, *Unto This Last*, *Works*, 17.25ff.

23. AH: Paw Hawk, *NM*, January 1898; "The Cats' Christmas Party," *NM*, September 1898; "The Customs of Cat Land," *passim*, *NM*; clothing, *NM*, August 1901; education, *NM*, February 1902; warfare, *NM*, May 1902; eating, *NM*, December 1901; poultry, *NM*, May 1899; three-legged lamb, *NM*, vol. 3, n.m. 1899. Gentleness to creatures, *Works*, 27.143; Teresa Smith notes that Robin invented "Jipandland" several years before *Nothing Much*; see Teresa Smith in R. G. Collingwood, *R. G. Collingwood*, 190.

24. "Two and a Night Line," *What Ho!* August 1905. She is staying at the Waterhead Inn. To be fair, the main character is also on holiday, but he clearly knows boats. Not surprisingly, sailing became a lifelong passion for Robin; later in life, he wrote of his experience sailing around Greece in R. G. Collingwood, *The First Mate's Log: Of a Voyage to Greece in the Schooner Yacht "Fleur de Lys"* in 1939 (Bristol: Thoemmes, 2003). Trippers, *NM*, May 1902, AH.

25. John Ruskin, ed., Alexander Wedderburn and W. G. Collingwood, trans., *The Economist of Xenophon* (Kent: George Allen, 1876), 56–58.

26. AH: launch of Dora, *NM*, December 1900; not dull or stupid, WGC to Dora, February 23, 1916.

27. Best-equipped mind, R. G. Collingwood, *Essays in the Philosophy of Art*, 40. Stuart Eagles, *After Ruskin: the Social and Political Legacies of a Victorian Prophet, 1870–1920* (Oxford: Oxford University Press, 2011), 268; but see Jose Harris, "Ruskin and Social Reform," in Dinah Birch, *Ruskin and the Dawn of the Modern*, 7–33. Robin Collingwood, *The New Leviathan: Man, Civilization and Barbarism* (Oxford: Oxford University Press, 1999[1942]). Ursula, Dora, and Barbara's

occupations, Townend, *The Vikings and Victorian Lakeland*, 250. Ironically, Robin was ambivalent about the work of the National Trust (see *The First Mate's Log*, 32).

28. See Altounyan, *Aleppo*: mountains, 119–50; sailing, 159–60; Scouts, 130; boarding-school decision, 163–64; Lanehead, 171–81.

29. First meeting, Arthur Ransome, *Autobiography*, 80–81.

30. "adopted," Ransome, *Autobiography*, 91, 94–96.

31. Hawthorn tree, Ransome, *Autobiography*, 96.

32. Hugh Brogan, *The Life of Arthur Ransome* (London: Hamish Hamilton, 1985), 301–3; Christina Hardyment, *Arthur Ransome and Captain Flint's Trunk* (London: Frances Lincoln, 2006[1984]), 18–19, 30–35; popular success, compare Ransome, *Autobiography*, 92. Dora's daughter Brigit (WGC's granddaughter) lived in Coniston and became president of the Arthur Ransome Society. See Christina Hardyment, "Obituary: Brigit Sanders," *Independent*, November 17, 1999.

33. See Karen Welberry, "Arthur Ransome and the Conservation of the English Lakes," in *Wild Things: Children's Culture and Ecocriticism*, eds. Sidney Dobrin and Kenneth Kidd (Detroit: Wayne State University, 2004), 82–96. Welberry notes Ransome's "profound influence" on the Lake District and spells out how his "vision of the lakes as a 'playground' was hugely influential in 'pragmatic and populist' terms in the 1930s and 1940s" (82, 84).

CONCLUSION

1. *Works*: Agnes's cottage, 28.256; early sketch of guild, 27.95–96; butter-making, free from towns, 28.261–62; no machinery, 27.87; railroads, 28.247. Cf. Graham Macdonald, "The Politics of the Golden River: Ruskin on Environment and the Stationary State," *Environment and History*, vol. 18 (2012), 125–50.

2. Ibid.: Agnes's books, 28.256; Agnes's proposed education, 28.266.

3. Ibid., factory towns, 28.267; Agnes in service, 29.487; Hilton, *John Ruskin*, 602–3; *The Economist of Xenophon*, ed. John Ruskin, trans. Alexander Wedderburn and W. G. Collingwood (Kent: George Allen, 1876), ix, xvii. For a few hints regarding the adult life of Agnes Stalker, see Viljoen, *Brantwood Diary*, 607–8.

4. Stuart Eagles, *After Ruskin: The Social and Political Legacies of a Victorian Prophet 1870–1920* (Oxford: Oxford University Press, 2011); Donella Meadows, Jorgen Randers, and Dennis Meadows, *Limits to Growth: The 30-Year Update* (White River Junction, VT: Chelsea Green, 2004); Paul Sabin, *The Bet: Paul Ehrlich, Julian Simon, and Our Gamble over Earth's Future* (New Haven, CT: Yale University Press, 2013); Will Steffen, Paul J. Crutzen, and John R. McNeill, "The Anthropocene: Are Humans Now Overwhelming the Great Forces of Nature?" *AMBIO*, vol. 36, no. 8 (2007), 614–21; Johan Rockström et al., "Planetary Boundaries: Exploring the Safe Operating Space for Humanity," *Ecology and Society*, vol. 14, no. 2 (2009), 1–33.

5. Bill McKibben, *The End of Nature* (New York: Random House, 2006[1989]); David Archer, *The Long Thaw: How Humans Are Changing the Next 100,000 Years of Earth's Climate* (Princeton, NJ: Princeton University Press, 2009), 19.

6. Bill McKibben, *Eaarth*, 102, 151.

Index

Page numbers in italic refer to figures.

advertisements and advertising, 19, 59, 100, 105, 116, 151, 160-63, 202n7
agriculture, 10, 17, 29, 36, 43-44, 66, 99, 140, 142, 154, 175, 176, 179
Altounyan, Taqui, 150, 171
Ambleside, 54, 97, 100, 101, 102, 103, 104, 117
Anthropocene, the, 177
anti-urban sentiment, 25, 27-28, 107, 120, 137, 175
apocalypse, 11, 14, 34-38, 98, 115, 144-45, 156, 178
archaeology, 19, 23, 55, 105, 119, 121, 125, 130, 131, 133, 149, 150, 154, 159, 164, *165*
Arrhenius, Svante, 177

Barrow-in-Furness, 42, 89, 145-46, 156
beauty, 2, 10, 12, 14-17, 23, 28, 32, 34, 37, 41, 49, 56-57, 64-65, 73, 75, 87-88, 96-99, 102-3, 105, 108, 126, 134, 141, 156, 166, 169, 172, 176
beekeeping, 175
Beever, Susanna, 18, 55, 62, 68, 70-95, 122, 127, 133, 134, 164, 167, 176, 177, 179; animals, 77, 167; artistic endeavors, 74, 80; botany, 73, 75, 78; childhood, 72, 74; food, 70, 75, 79, 82; *Frondes Agrestes*, 82-85, 88, 91, 92, 133; gardening, 70-71, 75, 76, 77, 78, 79, 82, 84, 85, 88, 89, 91, 92-94; *Hortus Inclusus*, 18, *71*, 90, 91, 92, 94; literary aspirations, 73, 74; natural history, *81*, 91; philanthropy, 74, 80, 87-88; queen, 72, 87-88; spinning, 80; Storm Cloud, 72, 88-92; story-telling, 82; Thwaite, the, 18, 70-73, *71*, 75-77, *76*, 79-80, 85-86, 91-94
Bible, the, 26, 36, 47
birds, 15, 77, 80, 82, 86, 94, 141, 156-57, *158*, 167
blasphemy, 36, 39, 41
bloomeries, 99, 146-47
book making: children's version of, 154, 164; handmade books, 3-7, *4*, *7*, 64-66, 154, 192n30
botany, 43, 73, 82
Brantwood, 6, 14, 17, 22, 24, 25, 29-30, *30*, 35, 42-48, 55-56, 68, 70, 73, 75, 77-79, 86, 90-92, 94, 98, 104, 118, 126-27, 129, 132-33, 150, 159, 173, 174-75

carbon dioxide, 116, 177, 189n30
cats, 82, 136, 164, 167-68
charcoal 45, 56, 98-99, 146, 159, 172
churches, 97, 103-5, 112-13, 117
climate change, 8, 10-11, 14, 17, 38-42, 72, 89, 91-92, 99, 177. *See also* Anthropocene; the Little Ice Age
coal, 3, 10, 12, 14, 19, 24, 33, 41-44, 50-51, 75, 88, 93, 99, 115-16, 144-46, 174, 176. *See also* pollution; scarcity

Coleridge, Hartley, 110–11
Collingwood, Barbara, 126, 132–33, 150, *151*, 152–54, 157, 160, 164, 169–70
Collingwood, Dora, 20, 126, 132–33, 142, 150, 152–56, *158*, 159–61, 164–65, 167, *168*, 170–72
Collingwood, Edith Mary "Dorrie," 120–25, 127, 129–30, 132–34, 150, *151*, 164, 171–72
Collingwood, Robin, 127, 133, 135, 150, *151*, 152–54, 156, 159–62, 164, 167, 169–70
Collingwood, Ursula 132, 153–54, 156–57, 164, 167, 170
Collingwood, William Gershom, 19–20, 29–30, *31*, 40, *41*, 54, 76, 85, 91, *93*, 105, 117, 119–48, 149–73 *passim*, 175, 176, 177; archaeology, 19, 23, 55, 105, 119, 121, 125, 130, 131, 133, 149, 150, 154, 159, 164, *165*; educational ideals (see *Nothing Much*); education at Oxford, 19, 29–30, 120, 122, 125; *Lake Counties, The*, 135, 139–40, 144, 146, 148, 156, 173; marriage, 120, 121–24, 126, 127; move to Lanehead, 127–29; *Philosophy of Ornament, The*, 120, 126, 141; preservationism, 121, 133, 142–43, 147–48; private secretary of Ruskin, 19, 40, 120, 125, 130, 134; Storm Cloud, 143–46; *Thorstein of the Mere*, 119, 120, 121, 135–40, *136*, *138*, *139*, 142, 159, 171, 173; tourism, 165–66, 169 (see also *The Lake Counties*); translates Xenophon, 29–31, *31*, 126, 169, 175; visit to Iceland, 140–43
Coniston (village), 44, 54–55, 68, 72–73, 86, 89, 92, 117–18, 125, 144, 147, 152, 156, 160, 171–73
Coniston Water, *16*, 89, 119, 135, 149, 156–57, 166, 173
conservation, 9–11
conservative, 47, 98
consumer society, 12, 14, 42, 78, 122, 151–52, 173, 176, 178
consumption. *See* ethics of consumption; sufficiency
coppicing, 45, 73, 147, 159, 174
cornucopianism, 11, 32, 177
cottages, 1, 6, 27, 54, 87, 100, 111, 118, 122, 150, 174–75, 178

Darwin, Charles, 2, 70, 77, 91
Darwinian, 46, 94

deforestation, 40–41, 146
division of labor, 7, 28
dyeing, 51–52, 60

education: art of living, 34; family magazine, 154; girls and women, 72, 79, 152, 202n5; Guild of St. George, 18, 50; home schooling, 152; landscape, 158; physical activity, 154; working class, 104, 175–76
Edward VII, 143
Elterwater, 55, 57
Enlightenment, 9–10, 14
environmentalism: anthropocene, 11–12, 177; conservationist, 9–11; popular, 11; preservationist, 9–10 (*see also* preservation); simple life and, 12
ethics of consumption, 1, 7–9, 12, 17, 23, 25, 32–33, 43, 48–49, 59, 63–64, 69, 74, 83–85, 88, 97–98, 112–13, 121–22, 137, 170, 176–77, 187n22. *See also* sufficiency

factory work, 2–3, 9, 26, 28, 48–49, 65, 115–16
fashion, 12, 56, 103–4
Fleming, Albert, 1–3, 5–7, 18, 55–69, 71, 77, 85–86, 90–1, 94, 105–6, 108–9, 113, 127, 154, 164, 176–77; conflict with Marian Twelves, 61–64; *Hortus Inclusus*, 18, 71, 90–92; *House of Rimmon, The*, 56–57, 61; learning to spin 1, 57–58; Neaum Crag, 6, 55; short stories 62, 191–92n24. *See also* Langdale Linen Industry
forecasting 8, 10–11, 19, 23–24, 33, 177
future: inheritance, 11, 92, 108, 113, 154, 158; technology, 114, 116; virtues, 8. *See also* forecasting

gardening: allotment, 17, 43; exotic plants, 85–86, 92; greenhouse (hothouse), 18, 44, 75, 78, 194n15, 195n27; moorland, 44, *45*; native plants, 18, 75, 194nn17–18; spade husbandry, 43; wilderness gardening (William Robinson), 75. *See also* Beever, Susanna
Garnett, Annie, 54, 106
geo-engineering, 11–12
geology, 10–11, 18, 77, 82, 147, 149, 177
glaciers, 9, 12, 14, 38–41, 147
God, 14, 17, 28, 36, 39, 80, 94, 149
Graham, John, 145–46
Grasmere, 54, 97, 100, 102, 104, 112

Guild of St. George, 18, 42–44, 50–53, 64, 79, 98, 101, 108, 145, 154, 174–75

Hill, Octavia, 87, 98, 104, 108, 145–46
Hills, W. H., 56, 59, 108, 114
household economy, 25, 30, 34, 44, 126, 128, 132, 137, 141, 175

Iceland, 140–43, 146
imagination, 19, 32, 47, 81, 122, 124, 135, 167, 170, 177. *See also* sufficiency

Jevons, William Stanley, 24, 33
Jura Mountains, 15, 17

Kendal, 107, 126, 130, 144, 150
Keswick, 54, 63, 99, 104, 107
Keswick School of Industrial Arts (KSIA), 18–19, 63–64, 68, 98, 105–8, *106*
Krakatoa, 35

lace, 57, 61, 67, 112
Lake District, the: geology, 99; history, 98–99; industry, 146–47; traditions, 64, 105, 112
Lake District Defence Society, 98, 107–8
Lake Geneva, 38, *41*, 144
landscape: as a book, 6; childhood influence, 18, 28, 73, 156, 158–59, 168, 172; destruction of, 97, 142; historic, 119, 135–36, 142; indifference to, 96; industrial, 115, 146–47, 159; labor and, 6, 49, 146; mountain, 14, 38; painting, 14–15, 122, 125; pastoral, 13, 15, 17, 25, 27, 98, 117–18, 146, 185n21; postindustrial, 19, 147; sacred, 17, 40, 158; shapes moral character, 100, 124, 158; sublime, 135; suburban, 27–28, 176; technology and, 117–18, 146; tourism, 100, 165, 173; tradition and, 17, 142; wilderness, 15, 146
Lanehead, 121, 127, *128*, 130, 132, 143, 149–50, 160, 170–73
Langdale, *5*, 55, 57, 66, 98, 144
Langdale Linen Industry, 18, 53, 55–68, *58*, 67, 80, 85, 105–6, 109
La Touche, Rose, 47, 74
Laxey Woolen Mill, 51–53, 59–61
liberalism, 28, 31–32, 36, 43, 47, 167, 170
limits to growth 10–11, 23, 31–32, 38, 44, 174, 176
Little Ice Age, the, 41

Malthus, T. R., 31
McKibben, Bill, 8, 177
Middle Ages, the, 28, 49, 54, 98, 167–68
middle class, 3, 12, 22, 27–28, 47, 56, 74, 79, 151, 175–76
Mill, John Stuart, 28, 31–32, 36, 38, 175
Millais, John Everett, 10
Mongoose Club, the, 154, 159–60
Mont Blanc, 38, 40, 144
Morris, William, 8, 49, 53, 108, 120, 149
Muir, John, 9, 14–15

National Trust, the, 13, 18–19, 87, 98, 103, 108, 148, 173, 176
natural history, 9, 14, 19, 23, 34, 83, 150, 154, 159, 173, 175
Norton, Charles Eliot, 38, 40, 70
Nothing Much (family magazine): animals, 156, 157, *158*, 164, 167–68; consumption (*see* advertisements and advertising); home schooling, 152–53; industrial history, 154, 156, 159, 160, 166, 173; literature and story-telling, 149, 152, 157, 158, 167, 169 (*see also* Ransome, Arthur); Mongoose Club and scouting, 154–55, 159, 160, 171; painting, 149, 150, 152, 153, *155*, 156, 158, *158*, 164, *168*, 169, 170; physically active learning, 153, 154, 156, 159, 172; Ruskin as "super-ancestor," 150; sailing and boats, 149, 150, 154, 157, 164, 165, 169–70, 171, 172, 173, 202n8, 203n24

Old Man, the (fell in Coniston), *16*, 25, 42, 55, 98, 121, 142, *143*, 149, 164, 172
Oxford, 26, 29–30, 40, 46, 73, 78, 90, 96, 104, 111, 120, 122, 125, 133, 170

pastoral, 99–101. *See also* landscape
paternalism, 62, 178
Peel Island, 121, 139, 149, 164–65, *165*, 171–72
Pepper, Elizabeth, 64–67, *67*
Plague Wind, the (a.k.a. Plague Cloud), 34–35, 39–40, 72, 89–90. *See also* Storm Cloud
planetary boundaries, 8
planetary degradation, 11, 14, 35, 37–39, 42, 176, 189n25
political economy, 28, 31
pollution: carbonic acid (carbon dioxide), 116, 189n30; smoke, 14, 35, 39, *41*, 43, 145,

pollution (*continued*)
    156, 166, 177; spiritual, 36, 120; water, 39, 52, 56, 89, 115, 145, 156
population growth, 9-11, 31, 98, 101, 166
preservation: landscape, 56, 64, 101, 107, 121, 148, 158; people and customs, 98, 101, 108, 140
Princen, Thomas, 8
printing, 2, 66, 154, 164
Promethean power, 37
prophecy (prophet), 36, 42, 101, 188n19, 189n24, 190n42

Queen, Ruskin's concept of, 72, 87-88

race: feline, 167; Lakeland, 100, 108-9
railways, 13, 18, 47, 56, 86, 96-97, 100-102, 107, 109, 117-18, 175
Ransome, Arthur: at Lanehead, 150, 171-72; *Swallows and Amazons*, 20, 121, 172-73
Rawnsley, Edith, 54, 68, 98, 103, 114
Rawnsley, Hardwicke, 13-14, 18-19, 54, 56, 63, 78, 96-118, 121, 127, 133, 134, 145, 146, 147, 177, 178; marriage, 98, 103-4, 114; research on Wordsworth, 109-12; Rushbearing Sermons, 112-13; smoke prevention, 115-17. *See also* Keswick School of Industrial Arts (KSIA); Thirlmere
Renaissance, the, 37, 47, 56
renewable energy, 12, 33, 44, 50-53, 80, 178
resilient nature: Big Blast, The, 160; *Vis Medicatrix*, 147; Romanticism, 9-10, 12, 14, 59, 73, 114. *See also* Wordsworth, William
Rushbearing (ceremony and sermons) 112-13
Ruskin, John: childhood, 25-26, 73; *Ethics of Dust, The*, 85, 152; *Fors Clavigera*, 14, 18, 19, 23, 37, 40, 43, 49, 60, 75, 83, 89, 96, 117, 174, 176; friendship with Susanna Beever (*see* Beever, Susanna); Hinksey dig, 29, 97, 104, 125; mental illness, 14, 18, 21-22, 37, 42, 47, 67-68, 71, 75, 89, 91, 92, 98, 144, 150, 177; *Modern Painters*, 6, 26, 33, 37, 72, 82, 83, 84, 85, 176; move to Brantwood, 29, 37, 42, 75; opposition to railways, 13, 47, 56, 96-97, 98, 100, 101, 102, 107, 109, 117, 175, 178; Oxford professorship, 19, 22, 26, 29, 47, 90, 125; patronage of Arts and Crafts, 51, 57, 61, 63, 64; photographs of, *13, 24*; *Seven Lamps of Architecture*, 15; *Stones of Venice, The*, 28, 48, 49, 83, 88, 176; Storm Cloud, the (*see* Storm Cloud); *Unto this Last*, 14, 23, 28, 33, 69, 74, 83, 85, 176, 179, 187n17; visits to the Alps, 144, 201n41 (*see also* glaciers)
Rydings, Egbert, 51-53

saga literature, 119, 136-37, 140-42, 152
sailing, 169-71
scarcity, 8, 11, 24, 33
Schor, Juliet, 8
Schumacher, E. F., 8
science, 9, 14, 36, 39, 46, 97, 116, 146. *See also* botany; Darwin, Charles; Darwinian; geology; glaciers; natural history
Scotland, 14, 25, 100
Scott, Sir Walter, 26, 76, 83
self-sufficiency. *See* sufficiency
servants, 21-22, 24, 67, 77-78, 91, 132, 161, 164
Severn, Joan, 22, 24, 31, 63, 66, 68, 92, *93*
sheep farming, 99, 111, 113
simple life, 2, 8, 9, 12, 17, 19, 24, 34, 72, 76, 78-79, 81-82, 84, 91, 95, 100-101, 109-10, 112-13, 119-20, 122, 126, 129, 140-41, 150, 169, 173, 175. *See also* sufficiency
Simpson, Arthur, 121, 126-31
slate quarries, 73, 107, *143*, 160
Smith, Adam, 26, 28, 36
smoke abatement, 116, 145-46, 201n46
socialism, 34, 49
spinsters, 71-72, 87
Stalker, Agnes, *45*, 174-76
state, 10, 29, 47, 170
statesman-farmer 140, 173
stationary state, the, 31, 38, 174-77
steam engine 2-3, 12, 33, 50, 98, 118, 140, 174, 178. *See also* railways
Storm Cloud, 9, 14, 34-42, 46, 75, 88-89, 97, 115, 144-45, 156, 170, 173, 176-77. *See also* Plague Wind
sublime, the, 41, 73, 135, 187n22
suburbs, 26-28, 100, 120, 175-76
sufficiency: defined, 8-9, 22-23, 83-84, 185n8, 186n10; practice of, 12, 17-19, 50, 72, 78, 80, 85, 150, 165-66, 173; redefining wealth, 17, 20, 25, 32-34, 43, 56, 78, 104, 113, 117, 126, 137, 142, 169, 176-77; self-restraint, 22, 34, 126, 153, 166, 169; self-sufficiency, 36, 42-44, 50, 99-100, 107, 119, 140, 173, 175; vicarious, 98, 178. *See also* imagination
Switzerland, 14, 39-41, 140

technology, 8, 23, 46, 66, 96, 115–18, 147–48, 160, 174, 178
Thirlmere, 13, 102–4, 107–8, 113–15, 134–35, 147, 176
Thoreau, Henry David, 8, 77
Thwaite, the. *See* Beever, Susanna
Tuckwell, William, 82, 86, 91, 92, 94
Turner, J. M. W., 15, 26, 35, 37, 82
Twelves, Marian, 57–58, 61–66, 68–69, 85, 106, 109
Tyndall, John, 39, 189n30

Venice 10, 37–38, 48, 54
Victoria, Queen, 162–63
vikings: on Iceland, 140, 142, 187; in the Lake District, 98, 112, 114, 119, 124, 135–40, *136*, *138*, *139*, 159

war: Cold War, 11; Franco-Prussian War, 38; Napoleonic Wars, 37; World War One, 24, 112, 150, 170; World War Two, 11, 54, 108
Warner, H. H., *Songs of the Spindle & Legends of the Loom*, 3–7, *4*, *5*, 7, 64–66, 68, 154, 192n31
waste, 3, 12, 23, 27, 29–30, 32–33, 96–97, 116, 147
water supply, 13, 29, 101–3, 134. *See also* Thirlmere
Wetherlam, 144
wilderness, 9–10, 15–17, 38, 75, 98, 135, 146–47, 187n21
Windermere 97, 104, 114, 122–23, 126–27, 172
women's work, 1–3, 6, 7, 51, 57, 62, 69, 72, 79, 87, 106, 109–10, 137. *See also* Garnett, Annie; Twelves, Marian
wood carving, 54, 56, 64, 80, 104–5, 127
Wordsworth, Dorothy, 110
Wordsworth, Gordon, 133, 143
Wordsworth, William, 99, 109, 111

Yosemite (Half Dome), 15

Xenophon, 29–31, 47, 49, 126, 141, 169, 175